Global Garbage

Global Garbage examines the ways in which garbage, in its diverse forms, is being produced, managed, experienced, imagined, circulated, concealed, and aestheticized in contemporary urban environments and across different creative and cultural practices. The book explores the increasingly complex relationship between globalization and garbage in locations such as Beirut, Detroit, Hong Kong, London, Los Angeles, Manchester, Naples, Paris, Rio de Janeiro, and Tehran. In particular, the book examines how, and under what conditions, contemporary imaginaries of excess, waste, and abandonment perpetuate – but also sometimes counter – the imbalances of power that are frequently associated with the global metropolitan condition. This interdisciplinary collection will appeal to the fields of anthropology, architecture, film and media studies, geography, urban studies, sociology, and cultural analysis.

Christoph Lindner is Professor of Media and Culture at the University of Amsterdam.

Miriam Meissner is Lecturer in Media and Cultural Studies at Lancaster University.

Routledge Research in Sustainable Urbanism

This series offers a forum for original and innovative research that engages with key debates and concepts in the field. Titles within the series range from empirical investigations to theoretical engagements, offering international perspectives and multidisciplinary dialogues across the social sciences.

Published

Co-producing Knowledge for Sustainable Cities
Joining forces for change
Edited by Merritt Polk

Global Garbage
Urban imaginaries of waste, excess, and abandonment
Edited by Christoph Lindner and Miriam Meissner

Global Garbage

Urban imaginaries of waste, excess, and
abandonment

**Edited by Christoph Lindner and
Miriam Meissner**

Routledge
Taylor & Francis Group

LONDON AND NEW YORK

First published 2016 by Routledge

2 Park Square, Milton Park, Abingdon, Oxfordshire OX14 4RN
711 Third Avenue, New York, NY 10017

Routledge is an imprint of the Taylor & Francis Group, an informa business

First issued in paperback 2018

British Library Cataloguing in Publication Data
A catalogue record for this book is available from the British Library

Library of Congress Cataloging in Publication Data
Global garbage : urban imaginaries of waste, excess, and abandonment /
edited by Christoph Lindner and Miriam Meissner.
   pages cm
   Includes bibliographical references and index.
   1. Refuse and refuse disposal. I. Lindner, Christoph, 1971– editor.
   II. Meissner, Miriam, editor.
   TD791.G56 2016
   363.72'88–dc23   2015026433

ISBN: 978-1-138-84139-0 (hbk)
ISBN: 978-1-138-54645-5 (pbk)

Typeset in Times New Roman
by Wearset Ltd, Boldon, Tyne and Wear

Contents

Figures

Contributors

Anne Berg is a Lecturer in the Department of History at the University of Michigan.

Anneke Coppoolse is a PhD candidate in the Department of Cultural Studies at Lingnan University.

Pedram Dibazar is a PhD Fellow at the Amsterdam School for Cultural Analysis, University of Amsterdam.

Nick Dines is a Research Fellow in the Department of Criminology and Sociology at Middlesex University, London.

Pauline Goul is a PhD candidate in French Literature at the Department of Romance Studies at Cornell University.

Geoffrey Kantaris is Reader in Latin American Culture at the University of Cambridge.

Christoph Lindner is Professor of Media and Culture at the University of Amsterdam.

Niall Martin is Lecturer in Literary and Cultural Analysis at the University of Amsterdam.

Miriam Meissner is Lecturer in Media and Cultural Studies at Lancaster University.

Judith Naeff is a PhD Fellow at the Amsterdam School for Cultural Analysis, University of Amsterdam.

Stephanie Newell is Professor of English and Senior Research Fellow in International and Area Studies at Yale University.

Brian Rosa is Assistant Professor of Urban Studies (Queens College) and Earth and Environmental Sciences (The Graduate Center) at the City University of New York.

Kirsten Seale is a Lecturer in Interdisciplinary Design at the University of Technology Sydney, and Adjunct Fellow at the Institute for Culture and Society, Western Sydney University.

Joshua Synenko is a Postdoctoral Researcher in Humanities at York University, Toronto.

Maite Zubiaurre is Professor of Spanish and Germanic Literatures at the University of California, Los Angeles.

Acknowledgements

This book has its intellectual roots in the ASCA Cities Project – a collaborative, interdisciplinary research initiative based at the Amsterdam School for Cultural Analysis. In the academic year 2013–2014, we organized a monthly research seminar at the Cities Project on the topic of 'Global Garbage'. During the seminar, we were struck by the extraordinarily high level of interest in garbage and the many ways in which the topic engaged scholars, students, and creative practitioners across disciplines in the humanities, social and environmental sciences, and design fields. Building on our seminar in Amsterdam, we next organized an international conference on the same topic, which took place in Paris in June 2014. This book registers and extends the discussions that took place in both Amsterdam and Paris. We would like to thank the Amsterdam School for Cultural Analysis, the Amsterdam Centre for Globalization Studies, and the University of London Institute in Paris for supporting this project. The participants of the Global Garbage research seminar and conference helped us to formulate the research questions, concepts, and arguments developed in this book. In this context, we are particularly thankful to Pedram Dibazar and Judith Naeff for their important intellectual collaboration in devising the project. Tijmen Klous provided outstanding research and editorial assistance. Most of all, we are grateful to the authors in this book for contributing their original work to this volume.

Christoph Lindner and Miriam Meissner
Amsterdam, 2015

1 Globalization, garbage, and the urban environment

Christoph Lindner and Miriam Meissner

Global cities and garbage

Garbage is increasingly central to contemporary globalization debates. An online search for the phrase 'global garbage' produces results such as 'global garbage crisis', 'global garbage detection', 'global garbage management', 'global garbage summit', and many more. Garbage has become a global concern. It is implicated in the transnational flow of goods, people, capital, data, and images that thinkers like Arjun Appadurai (1996) consider constitutive of globalization. However, in contrast to these forms of flow, garbage can also circulate globally in other, less obvious ways. For instance, garbage can circulate involuntarily, as alarming ecological reports on the many tons of plastic debris drifting around the world's oceans reveal (Derraik 2002). It can circulate for the sake of elimination, as in international garbage management and disposal programmes. Or it can even circulate as a commodity in its own right in a transnational 'second order market', where garbage is bought and sold for recycling or the extraction of raw materials.

Cities play a central role in this context. The average urban resident reportedly produces around four times as much solid waste as a person living in the countryside (Hoornweg *et al.* 2013: 616). Consequently, with more than half of the world's population already living in cities (United Nations 2014), urban population growth is expected to outpace waste reduction efforts in the near future (Hoornweg *et al.* 2013). For this very practical and urgent reason, cities also form key sites for experimentation with new strategies of waste management – as in the case of San Francisco's Zero Waste programme, which has the goal of sending 'nothing to landfill or incineration' by 2020 (San Francisco Environment, n.d.). Similarly, the concept of 'urban metabolism' has recently gained renewed attention in urban studies and planning. The urban metabolism approach seeks to apply a more holistic approach to urban garbage management, taking into account 'the sum total of the technical and socio-economic processes that occur in cities, resulting in growth, production of energy, and elimination of waste' (C. Kennedy 2007).

At the same time, due to elaborate recycling systems, specific types of garbage become potentially valuable commodities, whose collection and processing is predominantly carried out in urban environments. In *Junkyard Planet: Travels in the Billion-Dollar Trash Trade*, for instance, Adam Minter (2013) shows how the trade of specific waste materials constitutes a highly elaborate and extremely globalized market. In this worldwide market, waste materials are traded for the extraction of valuable scrap metals. Crucial points to make in this context are that contemporary global production and trading are coordinated via a network of what Saskia Sassen has termed 'global cities' (Sassen 1991), and that recycling has managerial urban nerve centres – such as the city of Shijiao, China, which processes around 20 million pounds of imported Christmas tree lights per year for the extraction of copper. This copper is then resold to neighbouring wire, power cord, and smartphone factories (Minter 2013: 1–2). Specific cities and regions are thus implicated in what happens before and after consumption.

Yet cities are more than just the main producers, managers, and marketplaces of waste materials. Garbage has also become prominent in urban social and artistic practices. Examples range from HA Schult's haunting armies of 'Trash People' (Figures 1.1 and 1.2) placed in remote locations in nature, as well as on central squares of prominent European cities such as Brussels, Cologne, Moscow, and Rome, to London-based street artist Francisco de Pájaro's 'Art is Trash' installations, made of urban detritus and assembled at different, seemingly random city sites.

Similarly, contemporary films, such as Lucy Walker's feature documentary *Waste Land* (2010), increasingly turn to the topic of urban garbage. *Waste Land*

Figure 1.1 'Roman People', 2007, installation by HA Schult (photograph by T. Hoepker. Courtesy of the artist).

Figure 1.2 'Arctic People', 2011, installation by HA Schult (photograph by G. Battista. Courtesy of the artist).

documents artist Vik Muniz's collaboration with *catadores* (pickers of recyclable materials) at Jardim Gramacho, one of the world's largest landfills near Rio de Janeiro. The documentary shows how Muniz and *catadores* assemble waste materials into artworks, which eventually get sold at a prestigious London-based auction house. Garbage is thus transformed into a commodity and circulates in the global art market. At the same time, Muniz's intervention is presented not only as an artistic project but also as a transnational social project, raising questions about global inequality and environmental justice. In this respect, *Waste Land* belongs to a broader tendency in contemporary culture to associate garbage – in its diverse forms and articulations – with issues of global economics, politics, and what social geographer David Harvey has termed 'uneven geographic development' (Harvey 2006). As the global flow of garbage can serve both the accumulation of profits (as in the trading of garbage for the extraction of raw materials) and the redistribution of risks, questions about where, how, and for what purposes garbage 'flows' are critical.

Beyond that, contemporary social and artistic engagements with garbage raise questions that evoke the Lefebvrian notion of a 'right to the city' – the demand for a 'transformed and renewed right to urban life' (Lefebvre 1996 [1968]: 159), to be realized in both practical and material ways. Garbage in this context becomes an allegory, whose characteristics stand for much broader problems and dynamics linked to globalization and its impact on urban development, such as the proliferation of urban slums, the acceleration of urban sprawl, the rise of transnational urban migration, and the privatization of public urban space and housing.

Five discourses on garbage

Responding to these issues – as well as to the broader role that cities play in garbage production, trading, debate, and reflection – this book develops an interdisciplinary, comparative, and transnational perspective on the relationship between garbage, globalization, and contemporary cities. In particular, it focuses on how that relationship is articulated in global media and urban culture. In so doing, the book engages and problematizes five critical discourses on garbage that tend to dominate existing scholarship.

The first of these comes from the urban planning and environmental studies perspective, which assesses environmental, social, and economic risks resulting from garbage proliferation, as well as the various possibilities of urban garbage management. A particular challenge that this perspective deals with is evaluating the global ecological impacts of garbage proliferation, while at the same time seeking to offer locally applicable solutions to garbage processing and reduction. Moreover, research on the environmental impacts of global garbage is confronted with the challenge of processing an extensive amount of largely unstructured data. As Zsuzsa Gille notes, 'there are no statistics on overall waste volumes. Instead, data are published on seemingly distinct categories, such as municipal waste, manufacturing waste' (Gille 2007: 15). International waste classification systems similarly vary and continuously change, which, according to Gille, 'also has to do something with the fact that the complexity of materials we produce intentionally or unintentionally increases faster than the capability of our classificatory systems' (Gille 2007: 17).

The second perspective on garbage and cities may be described as socio-anthropological. It analyses the socio-material practices as well as the attitudes that people – in particular urban inhabitants – develop in relation to different forms and sites of garbage. Socio-anthropological accounts of garbage practices and attitudes frequently work in the tradition of anthropologist Mary Douglas, whose assertion that dirt is 'matter out of place' – 'the by-product of a systematic ordering and classification of matter' (Douglas 2002 [1966]: 44) – inspires a critical reflection on garbage as the product of cultural systems of structuring and signification. Another focus of this perspective is on the socio-cultural environments of garbage creation and processing. Much research has been devoted, for instance, to the practice of garbage picking, stretching from scavenging as a subsistence strategy to dumpster diving as an activist means of consumer critique and subversion. Finally, the socio-anthropological study of garbage also comprises research on the role of cultural values, symbols, and education in relation to wasting and recycling in everyday life.

The third dominant perspective in contemporary scholarship on garbage analyses artists' use of waste material in works of assemblage and bricolage, as well as authors' and filmmakers' various engagements – both literal and symbolic – with trash. Crucial to such studies are the shifting cultural meanings that waste materials can acquire due to different artistic practices and art-historical paradigms. Gillian Whiteley has argued that 'with "the nomadic gathering of precarious materials and products" using "recycling (a method) and chaotic arrangement (an aesthetic)", the rag-picker and the bricoleur ... present powerful models for recent and current artistic practice' (Whiteley 2011: 8). The practice of garbage assemblage thus offers methods and aesthetics for contemporary artistic experimentation that have the potential to combine, restage, blur, or destabilize established cultural values and meanings.

In this way, artistic engagements with waste connect to a fourth perspective on garbage: the philosophical perspective. This fourth perspective engages the concept of garbage in relation to culture and ideology, including questions of metaphysics. The philosophical perspective both extends and problematizes the notion of garbage as 'matter out of place' by asking what qualifies garbage beyond the negative definition that it eludes established systems of classification and value. Beyond that, metaphysical questionings of garbage often prompt a critical questioning of the economic, symbolic, and epistemological systems that produce garbage in its various forms – such as trash, rubbish, residue, etc. Treating garbage as a lens through which Western culture and modernity are examined, John Scanlan argues that garbage 'utterances refer to the excrement of meaning itself. For example, it is when something means nothing *to you* that it becomes "filth," "shit," "rubbish," "garbage," and so on' (Scanlan 2005: 10). As the italicization of the words '*to you*' indicates, the designation of something as garbage thus also reveals semiotic and epistemological limits of the designating subject. Moreover, what Scanlan's focus on meaning ('means nothing *to you*') shows is that garbage may extend the sphere of material objects by referring to immaterial entities such as knowledge, language, and symbolism.

It is also in response to this last inflection that urban studies and geography have recently developed a fifth perspective on what may provisionally be termed 'spatial garbage'. Spatial garbage can be associated with terms such as 'badlands', 'blank spaces', 'derelict areas', 'No Man's Land', 'spaces of indeterminacy', 'terrain vague', 'urban deserts', 'vacant land', 'wasteland', and the list goes on. Critiquing contemporary discourses of urban planning, politics, and architecture concerned with these types of spaces, Gil M. Doron (2007a) has argued that these spaces are often integrated in some system of informal usage and valorization. Yet, in spite of the practical use of these spaces in everyday life, hegemonic discourses tend to devalue them.

It is partly in response to this dynamic that some branches of urban studies, as well as contemporary visual culture, have shown renewed interest in urban ruins. Caitlin DeSilvey and Tim Edensor, for example, have called attention to the way that 'we seem to be in the midst of a contemporary *Ruinenlust* [ruin obsession], which carries strange echoes of earlier obsessions with ruination and decay' (2013: 1). This *Ruinenlust* shifts between the celebration/romanticization of ruins as spaces of experimentation, subversion, and/or embodied spatial practices on the one hand, and dystopian portrayals of ruins as manifestations of crisis, decline, and destruction on the other. Yet, as DeSilvey and Edensor (2013: 15) conclude, 'the contemporary hunger for ruins transcends a simple romantic/ dystopic dichotomy, and speaks also to urgent desires to experience and conceive of space otherwise'. Both in urban theory and in the cultural imagination, 'spatial garbage' is thus treated ambiguously. This ambiguity is further complicated by the ways in which the notion of garbage is not only associated with spaces of ruination and abandonment, but also with the excesses of urban construction and development – such as the excesses of postmodern urbanism that architect Rem Koolhaas has conceptualized as 'junkspace' (Koolhaas 2002).

Of course, these various perspectives on garbage and cities overlap and inform each other. Yet they are also marked by considerable disconnections and paradoxes, which merit a more nuanced, interdisciplinary consideration. In *Junk: Art and the Politics of Trash* (2011), for example, Whiteley notes that, 'paradoxically, whilst the social outcasts and destitute children of India process lethal cyberjunk, in other parts of the world it is fashionable to work with trash' (Whiteley 2011: 6). In order to examine this paradox, it is important to go beyond its moral scolding (over Western escapism and a lack of global thinking) to develop a strong interdisciplinary critique of the various yet interconnected forms and meanings that garbage takes on in different global contexts.

In principle, this demand echoes Gille's observation that:

> Scholars studying waste in one form or another have been speaking at cross-purposes because they operate with different implicit definitions of waste. Especially unfortunate has been economists' assumption that waste is merely an attribute of efficiency, but public discourse has also been hampered by an environmentalist impulse to reduce the problem of waste to a problem of pollution.
>
> (Gille 2007: 14)

Gille's argument holds that cross-disciplinary perspectives can enrich the study of global garbage and help develop more productive approaches to dealing with garbage in its differentiated forms and localities. In this book, we seek to make a contribution to such a project by exploring how garbage is used, conceptualized, and imagined in relation to different urban contexts and cultures around the globe. To that end, in the next section we will take the concepts of 'global garbage' and 'urban imaginaries' as starting points for critical reflection, tracing how the chapters in this volume engage different aspects of the interrelation of globalization, garbage, and the urban environment.

Global garbage, urban imaginaries

In the tradition of Henri Lefebvre's (2009 [1974]) theory of the social production of space, critical urban studies have developed the concept of urban imaginary to refer to symbolic, cognitive, and discursive constructions of urban space and living. The idea is that such constructions determine not only the ways in which cities are conceived in, and structured through, urban planning and architecture, but also the ways in which people relate to, engage with, behave in, and interpret cities. Urban imaginaries are, for instance, constructed by means of artistic practice – such as literature, film, photography, and other forms of visual/textual culture – but also through everyday practices of language, communication, and street culture, as well as through top-down practices of urban design, development, policy, and place-making.

Crucially, the ways in which cities deal with garbage also contributes to the formation of urban imaginaries. The city of Guiyu in Southern China, for instance, has received much critical attention internationally because of its function as a global centre of electronic waste processing, and because of the dramatic health risks to which workers in Guiyu are exposed. Guiyu forms an extreme example, in that its global image is almost inseparable from its role in the global processing of e-waste.

Naples is another example of a city whose portrayal in the global news media often revolves around garbage and the alleged failures of urban waste management. Yet, as Nick Dines shows in 'Writing rubbish about Naples' (Chapter 8), which analyses the mediatization of Naples' urban refuse crisis in 2007, this image is often misconstrued, foreclosing a systemic political understanding of the city and its garbage problems. Central to Dines' argument is the insight that the way the city's problems are perceived and handled is shaped not only by its extended history of garbage crises, but also by its exceptionalist image as an 'extraordinary city'.

The role of city-specific local conditions in the politics of garbage is also the topic of Anne Berg's chapter on 'Waste streams and garbage publics in Los Angeles and Detroit' (Chapter 6). Comparing the garbage crises that hit Los Angeles and Detroit in the 1980s, Berg shows how the urban transformation of garbage processing practices and technologies is not just the result of global patterns but also reproduces local histories of public debate, citizenship, and social-racial inequality. Garbage, Berg argues, reproduces the systems that generate it.

From this analytical perspective, a first interdisciplinary conceptualization of global garbage takes shape. If garbage reflects societal structures and politics, it may be treated as a form of artefact – an artefact of globalization – reflecting and embodying the various forms of commodity use, exchange, consumption, and disposal that operate at global and local levels. A similar approach to garbage is taken in William Rathje and Cullen Murphy's *Rubbish! An Archeology of Garbage* (2001), which analyses garbage as a form of 'archaeology of the present', revealing global society's demographics and buying habits.

At the heart of this approach are two seemingly obvious yet noteworthy ideas. The first is that archaeological artefacts start out as garbage:

> An appreciation of the accomplishments of the first hominids became possible only after they began making stone tools, the debris from the production of which, along with the discarded tools themselves, are now probed

for their secrets with electron microscopes and displayed in museums not as garbage but as 'artifacts'.

(Rathje and Murphy 2001: 11)

The second key point is that if 'our garbage, in the eyes of the future, is destined to hold a key to the past, then surely it already holds a key to the present' (Rathje and Murphy 2001: 11). What is interesting about this second argument is that, if garbage can be considered an artefact waiting for an anticipated future archaeology, the study of 'garbage artefacts' could facilitate a somewhat distancing perspective on our present civilization. Garbage could thus be used as a means of defamiliarization – making familiar objects and practices newly strange – and would in this way allow for new, potentially innovative theorizations of taken-for-granted socio-cultural dynamics of the present.

The place of garbage in cultural theory and critique is also the subject of Geoffrey Kantaris' analysis of the films *Waste Land* (dir. Walker 2010) and *Estamira* (dir. Prado 2004) in his chapter 'Waste not, want not: Garbage and the philosopher of the dump' (Chapter 4). Yet, instead of applying an exclusively archaeological approach, the chapter follows waste picker Estamira, the protagonist of the homonymous film, who develops a philosophy of life based on the 'commodity stripped of its aura' – garbage. In so doing, as Kantaris elaborates, Estamira takes viewers into the space of the dump, thus subverting the distanced relations of vision that are dominant in contemporary portrayals of global garbage. Kantaris' analysis reveals the metaphysical potential of garbage, suggesting new insights into the enduring, complex nature of the commodity form and capitalist cultures of abstraction.

In this respect, Kantaris' chapter relates to another conceptualization of garbage that can be applied in interdisciplinary ways: garbage as commodity and/or the commodity's counterpart. As Kirsten Seale shows in her chapter, 'The paradox of waste: Rio de Janeiro's Praça XV Flea Market' (Chapter 5), waste is not merely the residue but in fact the material that sustains the commodity form. Beyond that, she argues, the question of whether waste remains residue or undergoes recommodification depends not only on the waste material itself but also on the aesthetic and symbolic orders into which that material is integrated, as in the case of Rio de Janeiro's Praça XV Flea Market.

An analysis of the visual orders of garbage's 'recommodification' is the subject of Anneke Coppoolse's chapter 'Under the spectacle: Viewing trash in the streets of Central, Hong Kong' (Chapter 10), which uses visual ethnography to explore Hong Kong's street culture of garbage picking. Following urban trash collector/merchant Shandong Lou, Coppoolse studies 'visual events' of collecting and organizing waste and its relation to the larger 'spectacle of flows' that characterizes Hong Kong as a global city.

The conceptualization of garbage as something integral to global systems of commodity production, valorization, and exchange also yields important insights into the ways in which contemporary cities are managed in the context urban branding and redevelopment practices. Exploring conflicting discourses that surround the former docklands of Pomona Island in Manchester, Brian Rosa shows in the chapter 'Waste and value in urban transformation: Reflections on a post-industrial "wasteland" in Manchester' (Chapter 12) how urban space is strategically devalorized as 'wasteland' to justify economically motivated real estate developments. Interestingly, this process ignores or downplays the various social, cultural, and ecological values that may actually be embedded in a space labelled 'wasteland'. Rosa therefore argues that the term wasteland obscures as much as it reveals, and that the value of urban space needs to be reconceptualized in ways that operate beyond the logic of capital accumulation.

One conclusion that can be drawn from this argument is that the conceptualization of garbage as commodity can provide a productive vantage point to question established value systems. However, it is important to note that, as Niall Martin argues in the chapter 'On Beckton Alp: Iain Sinclair, garbage, and "obscenery"' (Chapter 13), the treatment of garbage as something that offers a somewhat privileged epistemological and aesthetic perspective on the world draws on a tradition in European modernism extending from the work of artists and thinkers like Walter Benjamin, T.S. Eliot, and Marcel Duchamp.

In this context, two narratives are dominant: the narrative of garbage as relational (as an entity relative to something within a given socio-symbolic system) and the narrative of garbage as revenant (as refuse that refuses to be disposed of). Contrasting these two narratives, Martin analyses the portrayal of the slag heap at London's former Beckton Gas Works – Beckton Alp – in the work of British writer and filmmaker Iain Sinclair. Critically reflecting on Sinclair's Beckton Alp in relation to globalization and postcolonial memory culture, Martin argues that Beckton embodies a type of waste that refuses to be placed and, in so doing, also refuses to work for the production of subject/object distinctions. Accordingly, garbage acts as a figure of radical indeterminacy – a figure that may hold the power to destabilize hegemonic orders in fundamental ways.

Yet, this does not imply that figures of garbage always operate in anti-hegemonic ways. In the chapter 'Dirty familiars: Colonial encounters in African cities' (Chapter 3), Stephanie Newell shows that, in late-nineteenth and early twentieth-century African cities, the designation of African natives as 'dirty' served to impose colonial regimes of urban sanitation and racial segregation. Analysing journals by imperial-era travellers and traders, Newell identifies how the rhetoric of dirt establishes discourses of distinction that underpin colonial power relations. Putting a new twist on Mary Douglas' argument that dirt is 'matter out of place', Newell suggests that, in this particular context, dirt 'puts matter firmly into place'. In the process, Newell's analysis identifies a tendency rooted in colonialism to justify anti-cosmopolitanism via urban imaginaries of dirt. As the chapter further demonstrates, this tendency continues to inform contemporary public discourses about migration and mobility between global cities – such as, for instance, in the media coverage of the 2014/2015 Ebola virus outbreak within and beyond West African cities. Newell's analysis is particularly relevant to the conceptualization of garbage because it illustrates how urban imaginaries of garbage can act as tools of ideology.

Approaching garbage from the perspective of ethics in addition to ideology, Guy Hawkins notes in *The Ethics of Waste: How We Relate to Rubbish*, waste 'doesn't just threaten the self in the horror of abjection, it also constitutes the self in the habits and embodied practices through which we decide what is connected to us and what isn't' (Hawkins 2006: 4). Garbage as an ethic is mediated here via the body, through habits and disciplines. These physical practices are subject to continuous transformations, which are informed by much broader ideas about globalization, economic productivity, environmental responsibility, and more. According to Hawkins, this correlation is due to contemporary environmentalism's slogan 'think globally, act locally'. Thus, everyday practices of 'choosing a paper bag rather than a plastic one, composting, recycling, all indicate important shifts in our relationship to waste matter, how we manage it, and how guilty or righteous it can make us feel' (Hawkins 2006: 5).

Cities form central platforms and testing grounds where such ethico-ideological relationships between garbage and globalization are debated and negotiated. In her chapter 'Dirt poor/filthy rich: Urban garbage from Radiant City to abstention' (Chapter 9), Pauline Goul demonstrates how garbage ideologies – assumptions about the best systems of garbage production, avoidance, and discharge – become deeply implicated in modern urban thinking, planning, design, and practice. Moving from Le Corbusier's Radiant City, to contemporary domestic zero waste initiatives, and to the depiction of everyday garbage practices in contemporary film and documentary, Goul critically analyses the language, concepts, and aesthetics of garbage involved in these different visions of modern living. In particular, she reveals a historical tendency to associate garbage with fierce social stigma, and to tackle it with increasingly extreme efforts at eradication.

Goul's chapter addresses the ways in which modern visions of garbage are often connected to ideologies about ecological well-being. Such ideologies, she points out, fail to register the systematic socio-economic imbalances in which today's global garbage production is implicated. In order to achieve a truly sustainable social order of production, exchange, and wasting, those imbalances would have to be tackled either before or simultaneously with the realization of eco-friendly waste systems.

Touching on this argument, philosopher Slavoj Žižek argues in the documentary *Examined Life* (2008, dir. Astra Taylor) that 'the way we approach ecological problematic is maybe the crucial field of ideology today'. Standing in front of a large pile of garbage, Žižek describes the efforts of contemporary societies to erase garbage as an ideological operation, geared towards the re-establishment of some illusory state of purity and natural order. In part due to the efficient removal of garbage from sight – which causes waste matter to disappear from everyday life – people disavow ecological problems.

According to Žižek, people rationally know about the various ecological disasters of our time – from global garbage to global warming. Yet, because of the relative invisibility of such disasters, they cannot imagine ecological catastrophes and thus remain reluctant to change behaviour. 'What we should do to confront properly the threat of ecological catastrophe', Žižek therefore concludes, is not all this new age stuff to break out of this technological manipulative world and to found our roots in nature but, on the contrary, to cut off even more these roots in nature.... We should become more artificial

(Žižek in Taylor 2008)

And, pointing at the garbage pile, Žižek adds 'True ecologist loves all this' [*sic*].

In the film, it is not completely clear whether Žižek's provocative suggestion to 'love' and live among garbage envisions a future in which large piles of garbage proliferate without being considered a problem anymore, or if the engagement with garbage is meant to incite profound societal change – change that does not only concern global eco-ethics, but global political economy as well. Yet, what becomes clear from Žižek's comments is that garbage also needs to be analysed as a subject of aesthetics.

To address the aesthetic dimension of garbage, it is helpful to turn to a notion of aesthetics that encompasses the very orders of sensing and making sense of reality. It is a notion of aesthetics best understood in Jacques Rancière's terms as the 'distribution of the sensible' (Rancière 2004: 7). Understood this way, aesthetics describes the ways in which perception and thought are structured through politics, and it can be used productively to question how urban imaginaries of garbage serve or subvert power regimes.

This line of thinking is developed further in Maite Zubiaurre's chapter 'Trashtopia: Global garbage/art in Francisco de Pájaro and Daniel Canogar' (Chapter 2), in which she discusses how waste is aestheticized in the work of street painter Francisco de Pájaro and media artist Daniel Canogar. Comparing and contextualizing these works in relation to the artists HA Schult, Chris Jordan, Vic Muniz, and Mierle Laderman Ukeles, as well as the installations 'My Dog Sighs', 'Filthy Luker', and the Trashcam Project, Zubiaurre examines how contemporary art complicates the common association between garbage and urban/ global dystopia. The chapter shows the diverse ways in which contemporary art uses garbage as both subject and material of aesthetic practice. Depending on how garbage is materially processed, placed, and made visible, contemporary artworks bring into focus different stages and dimensions of garbage's travel from individual waste products of intimate scale to a dystopian mass phenomenon of global scope. However, Zubiaurre concludes that, despite their diversity, garbage artworks generally share an inclination to mix utopian elements into their dystopian aesthetics and commentary.

In his chapter 'Leftover space, invisibility, and everyday life: Rooftops in Iran' (Chapter 7), Pedram Dibazar shifts the focus from visibility to invisibility while still retaining an emphasis on aesthetics. Through his analysis of rooftops in Iranian cities, he theorizes invisibility and visual ambiguity as a key aesthetic quality of leftover spaces. The chapter shows that, although these rooftops may be considered 'wasted' leftover spaces in terms of urban planning and economic value, it is precisely because of their ambiguous status and lack of visibility that they can be appropriated for political protest and contestation. Dibazar analyses the illicit practice of installing satellite dishes on rooftops, as well as nocturnal shouting from rooftops, as forms of political protest. Close reading the temporal and aesthetic logics that are employed in such tactics of rooftop appropriation, he reveals the everyday politics that reside in the indeterminacy of leftover space.

A different case study of supposedly 'wasted' architectural space – that of the ruin – is examined in Judith Naeff's chapter 'Disposable architecture – reinterpreting ruins in the age of globalization: the case of Beirut' (Chapter 14). The chapter expands and problematizes approaches that focus on the material, affective, and aesthetic qualities of ruins by additionally considering the politics of past and present that shape the form, cultural meaning, and future of contemporary ruin spaces. Focusing on contemporary Beirut, where civil-war ruins persist amidst fast-paced urban development financed by global capital, the chapter develops the concept of 'disposability' to read Beirut's ruins as spaces that resonate with the precarization of life in the era of globalization. Because Beirut's ruins stand for a permanent state of exposure to violence and potential erasure, Naeff argues that it is important to regard their disposability as an enduring condition rather than a temporal or historical anomaly.

Implicit in Naeff's analysis is an understanding of ruins as a type of spatial garbage whose form and meaning are shaped by past and present events, appropriations, and power structures at local and global scales. Naeff therefore also conceptualizes the ruin as a space that has several layers of meaning. From this, an additional way of conceptualizing garbage emerges: garbage as a palimpsest – a layered artefact whose form, meaning, and representation are subject to a multiplicity of interventions and interpretations.

In urban and globalization studies, the concept of palimpsest has gained much attention, partly due to Andreas Huyssen's *Present Pasts: Urban Palimpsests and the Politics of Memory* (2003), which describes how contemporary urban imaginaries are marked by the increasing re-evocation of the past through memory culture. Accordingly, today's focus on memory stands in contrast to nineteenth-century urbanism, in which the past mainly served to substantiate a future-oriented narrative of historical progress and modernization. Huyssen identifies several problems connected to this tendency, notably the paradox that 'memory discourses themselves partake in the detemporalizing processes that characterize a culture of consumption and obsolescence' – 'especially when the imagined past is sucked up into the timeless present of the all-pervasive virtual space of consumer culture' (Huyssen 2003: 10).

In the chapter 'Geospatial detritus: Mapping urban abandonment' (Chapter 11), Joshua Synenko shows how abandoned city sites are made accessible to global memory culture through the imaging technologies of geospatial mapping and virtual tourism. The chapter examines two case studies. The first is Michael Heizer's *City*, an enormous sculpture in the Nevada desert that – against the artist's wishes – has been made visible to the public via Google Earth's satellite images. The second case study is Gunkanjima Island in South Japan, a former industrial coal mining site that – following its appearance in the James Bond film *Skyfall* (Mendes 2012), and with the aid of Google Street View – has been transformed into a popular site of virtual mapping, tourism, and imagination. Synenko's analysis expands existing reflections on the contemporary obsession with urban ruins by accounting for both the risks and potentials that are connected to the highly individualized practice of experiencing urban abandonment virtually. The chapter further engages with the concept of garbage as palimpsest by taking into account the conditions and technologies of imagination that shape the ways in which 'spatial garbage' can act as a cultural memory.

Waste, excess, and abandonment

As we have outlined above, this book examines global garbage from a range of disciplinary perspectives and across different cultural and geographic contexts spanning Beirut, Detroit, Hong Kong, London, Los Angeles, Manchester, Naples, Paris, Rio de Janeiro, and Tehran. To organize the chapters in a way that not only reflects the book's intellectual and geographic diversity, but also identifies resonances and interactions between chapters, we have organized the book into three interrelating, thematized parts: waste, excess, and abandonment.

Part I – Waste – focuses on the literal and material dimensions of garbage, examining how social and power relations in the city are shaped by various forms of waste, becoming either contested or reaffirmed through both formal and informal everyday practices. The essays in this part analyse everyday dynamics of garbage collection, circulation, and dispute, identifying both complicity with, and resistance to, globalization and its impact on urban governance and urban development. Part II – Excess – explores urban manifestations and imaginaries of leftovers, disarray, accumulation, and (creative) destruction. Investigating both the spatio-temporal and symbolic characteristics of excess, the essays in this part show how garbage is implicated in contemporary global systems of overproduction, supermobility, and hypermediation. In addition, the essays reveal how the topic of garbage can spill into seemingly unrelated public debates and political conflicts, such as those surrounding gentrification and post-industrial urban renewal. Part III – Abandonment – investigates sites and imaginaries of urban emptiness and dereliction such as ruins and disused infrastructures. Contrary to conceptualizations that view abandonment as a state of stagnancy or deadlock, however, the essays in this part retrace how, by generating social, symbolic, and affective 'afterlives' (in particular through nostalgia) such sites of abandonment become appropriated into urban everyday life and global systems of value creation.

Working within and across the broad themes of waste, excess, and abandonment, this book sets out to show the many, sometimes surprising, and frequently alarming ways in which garbage is implicated in contemporary social, political, and artistic debates about globalization and the urban environment. Understanding garbage, we therefore argue, offers a way into understanding a number of urgent global issues, ranging from human health, social justice, and environmental risk, to uneven geographic development, neo-liberal capitalism, postcolonial politics, and transnational governance.

Part I
Waste

2 Trashtopia

Global garbage/art in Francisco de Pájaro and Daniel Canogar

Maite Zubiaurre

Utopia is a place foreign and remote, a point of destination and a reward after a long journey riddled with obstacles and difficulties, whereas Dystopia is usually the point of departure, the dreadful place one abhors and longs to leave behind. In the case of rubbish, however, the opposite is true. Waste, also an eager traveller (Hodding Carter 2006; George 2008; Hohn 2011; Nagle 2013; Rogers 2005; Royte 2005; Thomson 2009) and a global phenomenon, moves from 'utopia' (the 'happy' land of consumerism and affluence) to dystopia. In fact, trash carries dystopia within, is dystopia, particularly when allowed to grow and to pile up on dumps and landfills (Boo 2014; Engler 2004; Humes 2012; Rathje and Murphy 2001; Urrea 1996). In this chapter, I reflect upon refuse – and the journey that leads to (dystopic) monumentality – as represented in the work and installations of street painter Francisco de Pájaro, and media artist Daniel Canogar. The two Spanish artists do not work in isolation, as the following pages try to show, but dialogue with a number of international artists and installations, among them, HA Schult, Chris Jordan, Vic Muniz, Mierle Laderman Ukeles, 'My Dog Sighs', 'Filthy Luker', and the TrashCam Project.

The fact that trash travels, for example, is equally relevant to de Pájaro, Jordan, Canogar, and Schult, but the persistently nomadic nature of rubbish impacts the four artists differently. Jordan and Canogar look at trash as it arrives at transfer stations and recycling centres, and go through the sorting process, whereas Schult likes to conceive of trash as an eternal traveller and even an unwelcomed refugee, forced to move from place to place. De Pájaro, on the other hand, looks at rubbish well before it leaves the urban topography. The type of refuse that ignites his imagination is not the apocalyptic vastness and trash accumulation that happens in transfer stations, recycling centres, and landfills (as is the case with Jordan and Canogar), but the humble piece of rubbish that pauses briefly right at the beginning of its long journey, and provisionally litters sidewalks and street corners.

However, no matter how different their approach to trash, or the nature of their artistic manipulation of all things discarded, de Pájaro and Canogar, no less than Jordan and Schult, effectively complicate the routine perception of trash as a dystopic reality. They also are quick to acknowledge that, independently from the fact that the meaning of trash changes when art comes into play, in the 'real world' trash means, and is, very much the same, wherever it goes, and wherever one finds it. It is a global phenomenon – as is the cycle of production and consumerism – and the sameness/globalism of it also plays an important role in de Pájaro's and Canogar's approaches to the discarded.

'*El arte es basura*' (art is trash) is Francisco de Pájaro's signature, which he scribbles, graffiti-like, on the pieces he creates with the help of trash bags and old cardboard boxes. His art pieces were first seen in the crowded streets in Barcelona, but later also on the busy pavements of London and New York. De Pájaro, an artist from Extremadura, Spain, soon became deeply disenchanted with the exclusionary elitism of the art world. As he pointed out in a recent interview,

> I started painting on the street because the art galleries of Barcelona closed [their] doors [on me]; it was very difficult to evolve with limited financial means. Expressing myself on the trash gives me endless elements to paint, without having the obligation to maintain or market the finished work. You paint it, leave it in the street and keep going until you get tired. There's nothing like painting on the streets for freedom of expression.
>
> (Albanese 2014)

The fact thus that the established art world is exclusionary, that art materials are expensive, and that his meagre financial means could not keep up with his effervescent and quick-paced imagination, compelled de Pájaro to look for cheaper canvases. And ... what could be cheaper than trash bags? In fact, the shiny (mostly black or white) surface of plastic, and the bloated contours of bags bursting with rubbish not only was free for grabs (and, in de Pájaro's eyes, thirsty for colour), but also offered an invaluable opportunity for the public display of his talent. Thus armed with a skimpy rucksack that contained a few essentials – a couple of cheap brushes, four or five colour tubes and sprays, a pair of scissors, and some glue – de Pájaro took by storm the streets of nocturnal Barcelona. His working method was simple, and also pedantically fixed on routine. He would splash colour with broad strokes on trash – plastic bags, discarded furniture, cardboard boxes left at kerbsides, etc. – four nights a week (the other three he worked as a waiter in one of Barcelona's most venerable restaurants), and he would do so between 7 PM (which is when trash cans were rolled into the sidewalks, and bags and boxes would start littering the sidewalks), and 10 PM, the time of the arrival of municipal trash collectors. De Pájaro worked feverishly during these three hours, quickly pacing the streets on the lookout for trash – or clusters of trash – that appealed to his imagination, concentrating solely on the task at hand, oblivious to what was happening around him, and never answering questions from curious pedestrians.

I met de Pájaro back in the summer of 2012, and spent a whole night with him roaming the streets of Barcelona. I did not have a chance to see de Pájaro's art work that particular night, because, as stated above, de Pájaro refuses to talk and paint at the same time. Also, his pieces only last a few hours, the short span between birth and death, between the making of the piece and the arrival of the garbage truck. The provisional nature and ultimate fragility of his trash pieces are very important to de Pájaro. In fact, it infuriates him when passers-by try to 'rescue' his pieces and take them home. If we have pictures of some if his early artwork, it is because a friend started following him with his camera, keen on preserving a lasting copy of short-lived creation.

Since then, de Pájaro's art, which was essentially unknown beyond the Barcelona street art scene, has grown exponentially, and has materialized in other cities and continents (London, New York). Francisco de Pájaro is now an artist whose fame is escalating (a London gallery recently housed an exhibition of his work), and who nonetheless remains proudly faithful to his early artistic agenda, his *modus operandi* (thus far, de Pájaro has not traded his cherished trash bags for 'real' canvases), and his ideological principles. For de Pájaro, 'art is trash' (*el arte es basura*), but the reverse – 'trash is art' (*la basura es arte*) is equally true, and probably more essential to his work and to the cultural and pictorial tradition that his art (consciously, or unconsciously) embraces. De Pájaro's opus is *art engagée*, in its purest – and perhaps quite old-fashioned – form. For once, his artistic creations connect with the deep-seated Romantic tradition of 'animating the inanimate', of which the literary piece by German Romantic poet Heinrich von Kleist, *Ueber das Marionettentheater* (1810) is an illustrious early example, and the film *Toy Story* (1995), a very recent and no less famous one. In fact, the 'animation principle' also informs street/trash art, with an even more direct link to de Pájaro's oeuvre. Two British street artists/activists, known by their pseudonyms, 'My Dog Sighs' and 'Filthy Luker', have set out recently to paint eyes on discarded objects on sidewalks and at street corners and intersections, which intently look at the passers-by.

An even more compelling and sophisticated example is the *TrashCam Project*, a project created and developed in Hamburg, Germany, that instils life into trash dumpsters, and endows them with eyes that see and photograph:

> Four German sanitation workers and one photographer put their skills together for an unconventional art endeavor called the Trashcam Project, turning 1,000 liter dumpsters into giant, makeshift cameras. Fitting the bins with pinhole cameras, the group toured their favorite spots in Hamburg, capturing stunning black-and-white photographs of the city.
>
> (Genuske 2012)

'Maintenance Artist' Mierle Laderman Ukeles's artistic intervention, 'Mirror Truck', constitutes a similar experiment, where inanimate objects of the most 'abject' kind (trash, or trash-related) become alive, and 'dare' to look at their surroundings. Since 1983, the 'Mirror Truck' drives along during parades and community festivals, a burlesque reincarnation of Stendhal's famous dictum, 'a novel is a mirror walking down a road'. It certainly embraces much of Stendhal's romantic spirit (Stendhal never fully endorsed or represented hardcore realism), and much of the Romantic fervour of his fellowman Victor Hugo.

Sanitation workers are the *Miserables*, and the harsh reality of the city (be it Paris, or New York) is reflected in their eyes, and in the big, square 'mirrors/spectacles' of Ukeles's garbage truck.

Francisco de Pájaro thus is yet another of the trash *provocateurs* who have dared to endow garbage with the sense of sight. And it is through the close scrutiny of some of his street art pieces (Figures 2.1, 2.3, 2.4, and 2.5) that we

Figure 2.1 Francisco de Pájaro, *Untitled*, Madrid, 2012 (courtesy of the artist).

Figure 2.2 Francisco de Pájaro, *Untitled*, New York City, 2014 (courtesy of the artist).

Figure 2.3 Francisco de Pájaro, *El Arte es Basura no abandona a la Indigencia*, Barcelona, 2011 (courtesy of the artist).

Figure 2.4 Francisco de Pájaro, *Untitled*, Barcelona, 2012 (courtesy of the artist).

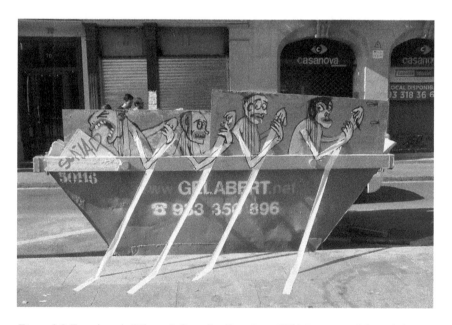

Figure 2.5 Francisco de Pájaro, *Indignados*, Barcelona, 2011 (courtesy of the artist).

will be able to answer the question of what happens when, instead of being us who stare at trash, it is trash that stares at us. First, it is important to note that, in my conversations with Francisco de Pájaro, he insisted on how boring trash is, how fixed it is on routine, with an arrival and departure schedule that makes the German railway system look sloppy. Mostly, though, trash is always the same, tediously so, according to de Pájaro:

> 'La basura es siempre lo mismo', 'la gente siempre tira las mismas cosas, las bolsas llenas de resto de comida, las sillas cojas, los colchones viejos y manchados. Muchas sillas, y muchos colchones.' (Trash is always the same. People always throw away the same stuff, plastic bags full of food remains, limping chairs, and stained mattresses. A lot of chairs, and a lot of mattresses.)

And it is precisely this monotone 'sameness' of discarded objects that appeals to de Pájaro's critical eye and his strong political consciousness, for in the dull uniformity of refuse he identifies the numbing unvariedness of the urban masses. Or rather, the way in which the latter is perceived by the powerful and affluent few. De Pájaro's effort at imprinting the discarded objects with human-like traits that make them different from each other is thus an attempt at humanizing the urban crowd, and at endowing it with individualized faces and souls.

De Pájaro's trash art aims its incisive darts at Baudelaire's dandyesque voyeur in *The Painter of Modern Life*, and only partly identifies with him. For de Pájaro is not as much interested in looking, or at being looked at (that, again, would turn him into yet another predictable *flâneur* and *fin de siècle* dandy), as he is in bestowing the power of sight on the disenfranchised. De Pájaro's distorted trash creatures have expressionistically eloquent faces, with wide-open eyes that defiantly reclaim the attention of pedestrians. They flatly refuse to be overlooked, the way trash (and the poor and homeless) usually are. Moreover, they demand from us that we listen to their stories, because de Pájaro's trash creations have a marked theatrical quality to them, and more often than not turn into *tableaux vivants* imbued with harsh political content. De Pájaro's performative trash art dialogues with Augusto Boal's 'theater of the oppressed', and bears an even stronger resemblance with Francisco de Goya's *Caprichos* and *Desastres de la Guerra*. Like his illustrious namesake, de Pájaro is heavily invested in painting the lives of the poor and the wretched. And like Goya, he is fearless when it comes to denounce the horrors of poverty and oppression, and to give voice to the voiceless. Take the trash piece, for example, of a crippled homeless covering himself with a blanket as discarded as him (Figure 2.2), or, even more cunningly tragic, of a mother begging for food and holding on to her small children (Figure 2.1).

De Pájaro's trash art is stark and never fails to confront human misery head on. He is a friend of the abject, and more often than not caters to scatology and dark humour. Particularly among his early pieces, nakedness prevails, crass sex scenes are not uncommon under the Barcelonan sky, and discarded toilets often serve as prime material to bare buttocks. One of de Pájaro's trash creatures, for example, is depicted as defecating at a corner, while at the same time defiantly brandishing a stick. He presses his index finger against his lips, asking us – in jest, of course – to keep his smelly secret. But who is the irreverent jester? It is our trash. It is us. It is Diogenes of Sinope, the cynic, reminding us, through his own laughing obscenity and triumphant debasement, of our true human nature. In Francisco de Pájaro, Diogenes of Sinope and Saint Francis of Assisi unite, for de Pájaro's art is not only unabashedly cynical and irreverent, but also fiercely compassionate (Figure 2.3).

As I pointed out earlier, in all of his trash art pieces, de Pájaro characteristically and insolently signs with '*El arte es basura*' (art is trash). In Figure 2.3, however, his signature makes an unexpected sharp turn. The word 'basura' is missing, although if we look closer, we see it has just retreated to one of the sides of the box. In any case, the sentence seemingly ends with 'es' (is): there is a gap, or a pause, and then the sentence resumes below, with '*no te abandona*' (does not leave you). '*El arte no te abandona*' (Art does not leave you). Francisco de Pájaro, by way of his trash creature, is writing a consoling message to the homeless lying on the floor, for when the homeless wakes up.

De Pájaro, no less than trash, is an indefatigable traveller, and his *art engagée* has emerged from the debris of a number of cities, first Barcelona, and more recently London and New York. De Pájaro identifies trash as a globalized 'commodity' and universal language 'consumed' and spoken by natives and foreigners alike. But he also endows it with certain 'regionalist' or idiosyncratic features that directly hint at social and environmental issues specific to local culture. Figures 2.4 and 2.5 are cogent examples of de Pájaro's double-tongued aesthetics. The 'dumpster-canned trash-eating-trash fish' (Figure 2.4) is a straightforward critique of our polluted oceans, and a compassionate homage to the many species that become victims of trash-infested water. But it also makes particular sense in Barcelona, a coastal city with an active fishing industry. In the same wake, the rowing galley of grim-faced men/slaves toiling on a 'dumpstership' (Figure 2.5) speaks eloquently of the enslaving labour conditions of exploitative global capitalism, and brings back sombre memories of the longstanding Western tradition of slavery and oppression. But it also hints bluntly and unapologetically at the gruelling working conditions often imposed on African immigrants in contemporary Catalonia and other affluent enclaves of the Iberian peninsula.

It is important to note that Francisco de Pájaro is himself an immigrant (he migrated from a small town in Extremadura to Barcelona), and that his work is that of an artist who creates from the margins, and in defence of the marginalized. Thus wherever de Pájaro takes his art, he readily positions himself among the disenfranchised. And he once again resorts to his usual 'double' strategy of tackling global flaws and tragedies (such as poverty and social injustice and oppression), while at the same time pointing to specific and local versions of it. In Manhattan and on the American continent, for example, de Pájaro seamlessly transforms from '*charnego*' (a pejorative term for immigrants who flock into

Catalonia from Andalusia and Extremadura) into a defiant Native American, and not only metaphorically so. A number of pictures taken during his Manhattan stay depict de Pájaro dressed as feather-wearing redskin. Yet another set of pictures showcase his New York trash art, where the oppression of Native Americans, and of Native American's rejection of their colonizers and oppressors, become a frequent theme.[1] Asked during an interview about the underlying reasons of his frequent depiction of Native Americans on the canvas of Manhattan trash, de Pájaro responds as follows:

[Esas imágenes de indios] nacen de las mentiras de Hollywood. De niño, nunca me creí que los malos fuesen los indios en sus películas. También en la escuela, me enseñaron la falsedad de la Conquista de España como algo maravilloso y grandioso de nuestra historia. Yo he nacido en Extremadura, una región muy bonita que vio nacer a los sanguinarios más grandes de la Conquista del Continente Americano. También me siento identificado con la forma de vida, con el respeto y la filosofía que los nativos tenían. Tan solo tenemos que mirar a nuestro alrededor para darnos cuenta el daño tan grande que estamos causando a la naturaleza con nuestra forma de vivir. Mi espíritu es la de un guerrero que pinta para que no se nos olvide que todos somos hijos de la naturaleza, cada uno tenemos que hacer nuestra lucha interior para preservarla. Estoy alineado con todos los pueblos indígenas.

([The images of Native Americans] are born out of Hollywood's lies. As a child, I never believed that the bad guys in the movies were the Indians. At school, teachers also propagated a number of falsehoods about the Spanish Conquest as a wonderful and glorious event of our History. I was born in Extremadura, a beautiful region that was the cradle also of the most sanguinary conquistadors of the American continent. Also, I closely identify with the way of life, the respect, and the philosophy of Native Americans. We only need to look around to see how much harm we are inflicting on Nature with our way of life. My spirit is that of a warrior who paints so that we don't forget that we are children of nature, and that we all have to keep up our internal fight to preserve our natural world. I am aligned with all the native people.)

(Albanese 2014)

Francisco de Pájaro's passionate, and compassionate *art engagée* takes place on the streets of 'sanitary' cities in the Western hemisphere, where trash cans patiently wait on pavements, empty boxes pile neatly against trees, and dumpsters never really overflow. This artist's garbage is still somewhat measurable, somewhat 'domestic' and 'individualized'. It is thus relatively easy to humanize it, to create a soul for it and a face capable of expressing feelings, and of triggering emotional responses in others. But … what happens when garbage loses all its 'human' proportions, when countless computers, printers, televisions, VCRs, mobile phones, fax machines, stereos, electronic games, and wrecked cars – on top of the usual trash bags, cardboard boxes, stained mattresses, and mangled pieces of furniture – turn into an unstoppable avalanche? What happens when the gates of transfer stations, recycling centres, and landfills open and garbage floods in, with the force and violence of a tsunami? This is the experience – of trash that defies experience – artists such as Chris Jordan and Daniel Canogar portray. In his photographic series, dubbed *Intolerable Beauty: Portraits of American Mass Consumption* (2003–2005), Jordan looks precisely at rubbish gone immense, its inhumane proportions, ironically, brought upon nature by humans themselves. Here is what Jordan has to say about his series:

> Exploring around our country's shipping ports and industrial yards, where the accumulated detritus of our consumption is exposed to view like eroded layers in the Grand Canyon, I find evidence of a slow-motion apocalypse in progress. I am appalled by these scenes, and yet also drawn into them with awe and fascination. The immense scale of our consumption can appear desolate, macabre, oddly comical and ironic, and even darkly beautiful; for me its consistent feature is a staggering complexity.
>
> (Jordan 2005)

Jordan's photographic depictions of monumental waste as a tragic consequence of massive consumption are intriguingly monographic. Rather than portraying trash in the style of conventional landfill iconography – a towering conglomeration of rubbish where all things discarded mix and blend into each other – he chooses to carefully select and sort discarded items. Jordan's (and also Canogar's and even Schult's) *modus operandi* are very different from that of de Pájaro. The latter tackles trash at its most 'primitive', and before civilization and its heavily regulatory system have a chance to intervene further. As he states in an interview, '[I paint] as did our primitive ancestors, directly on the walls of caves […]. For me, painting is the mother of the arts, and [to paint] directly on any surface makes me a child of Nature' (Albanese 2014).

Jordan is much less of an *improvisateur*, and his ties to nature are less evident. His interest, rather, lies in the imprint of civilization on trash. Civilization creates trash, and also tries to impose an order on it, and it is this concept of waste carefully organized into categories that exerts a peculiar fascination, invariably mixed with horror, on artists like Jordan and Canogar. Jordan's approach in particular is that of a recycler/collector, who seems to derive a strong sense of accomplishment and power from throwing yet another item into the recycling bin, or adding it to his collection (Camille and Rifkin 2001; Baudrillard 1994a; Stewart 1994). Sure enough Jordan's *Intolerable Beauty* series showcases monothematic photographs of carefully separated items, such as discarded mobile phones, circuit boards, phone chargers, spent bullet casings, cigarette butts, diodes, broken glass, crushed cars, and even sawdust mountains.

Spanish media artist Daniel Canogar's series of photocollages, *Otras geologías* (*Other Geologies*), also first shown in 2005, builds on Jordan's view of trash as an alien landscape of vast geological proportions.[2] In Canogar's words, 'Nuestra basuras están creando un nuevo paisaje excremental que preferimos no ver, razón por la que se retira a la periferia de la ciudad' (Our trash is creating a new excremental landscape that we choose not to see, and that is why we are moving it to the outskirts of our cities) (Ibarz 2009). He also relates to the interest of his American counterpart in the classifying fervour of garbage operations around the globe:

> El [...] mundo de las basuras es muy especializado, en él se esta constantemente separando y reagrupando los materiales, hay todo un proceso de selección para intentar reciclar algunos materiales, otros no pueden ser reciclados por ello hay que enterrarlos directamente. En mis visitas a las cacharrerías industriales, a los puntos limpios, o a los centros de tratamiento orgánico he visto como se separa la basura por bloques o separadores. De alguna forma yo quería representar esos separadores de basura, en forma de cajones o bloques temáticos, para almacenar la basura conceptual.

> (The world of trash is a very specialized one, where materials are constantly separated and later reorganized into different groups. There is a whole selection process in place, in an effort to recover certain materials; the ones that cannot be recycled are discarded and buried. In my visits to recycling centres, industrial transfer stations, and organic reclamation plants I was able to witness how waste would be separated into blocks or bins. To a certain extent, I wanted to represent those blocks also by creating drawers or thematic blocks of conceptual trash.)

> (Ibarz 2009)

Canogar's junkscapes, however – so uncannily similar to Jordan's, so Grand Canyon-like, with its 'eroded layers of accumulated detritus' – add one unexpected ingredient to carefully sorted-out trash: the human body (Figures 2.6, 2.7, and 2.8). Canogar is just filling in the gap, adding the missing piece to the ultimately incomplete, and even 'unrealistic' depiction of junkscapes à la Jordan: For landfills are not lunar landscapes, eerie geologies devoid of any human presence, but 'real' and earthbound loci heavily infused with human presence and responsibility.

Wherever there is trash, Canogar's art implies, there are humans (both in the form of victims and perpetrators), an implication also present, and compellingly so, in the trash pieces of Francisco de Pájaro, not only because he consistently and stubbornly 'humanizes' and anthropomorphizes rubbish, but also because he faithfully records the brutal reality of trash and humans sharing urban filth. Trash and flesh equally blend into each other in Canogar's *Otras geologías*. Naked bodies of humans mix freely with the bodies of discarded computers (Figure 2.6); a human eye peeks at us – one more eye among innumerable jettisoned camera eyes; and the blonde mane of a woman cascades down, together with the dark mane of unwanted cassette reels (Figure 2.7).

Canogar's 'inhabited' garbage geologies invite at least three different readings, none of which preclude each other. The most obvious one, dear to

Figure 2.6 Daniel Canogar, *Other Geologies* (2007), 150 × 225 (courtesy of the artist).

Figure 2.7 Daniel Canogar, *Other Geologies* (2005), 150 × 225 (courtesy of the artist).

Figure 2.8 Daniel Canogar, *Other Geologies* (2007), 150 × 225 (courtesy of the artist).

catastrophizing environmentalist thought, is that trash will devour us, destroy our planet, and crash our ecological balance, in other words, that we will become the victims of our own consumerist greed. Jordan's reflections on his own project, *Intolerable Beauty*, seem to endorse such an interpretation: 'The pervasiveness of our consumerism', he says,

> holds a seductive kind of mob mentality. Collectively we are committing a vast and unsustainable act of taking, but we each are anonymous and no one is in charge or accountable for the consequences. I fear that in this process we are doing irreparable harm to our planet and to our individual spirits.
>
> (Jordan 2005)

A second reading, deeply embedded, incidentally, in Spanish cultural tradition and history, is that humans themselves are 'trash' among trash, flesh destined to perish and to decompose. The baroque passion for death symbols, and the obsessive prevalence of the *memento mori* theme resonates with Canogar, who has often acknowledged the influence that the Spanish and Italian baroque, particularly Rivera and Caravaggio, have exerted on his own art. Not only graveyards and skulls, but also landfills and stained mattresses, as well as electronic waste – Canogar often uses the term '*barroco electrónico*' (electronic baroque) – sombrely utter the words, 'Remember that you must die!' The 'plastic toy dump' of one of Canogar's 'trash collages' is a particularly compelling *memento mori*, where life and youth – as represented in the playful cheerfulness of toys, and the nude body of a pregnant woman (Figure 2.8) – clash sinisterly with death and decay: we know that the colourful plastic landscape is not a playpen, but a landfill/graveyard full of discarded toys, doomed to rot, like dead people.

The Spanish contemporary media artist relates modern excess – an excess of consumption that then degenerates into an excess of trash – to *horror vacui*, the fear of emptiness, yet another baroque theme. In fact, in one of his interviews, Canogar refers to the Diogenes Syndrome, the pathology of compulsive hoarders, as the paradigmatic illness of savage capitalism. But mostly, and this leads us to our third interpretation of *Otras geologías*, what propelled the Spanish artist to throw naked bodies into landscapes of excreted excess was the painful confirmation that humans not only share the destiny of things (they die and rot; they become trash), but are treated by others as disposable items. As Canogar tells us, he decided to populate *Otras geologías* with nude anatomies when he saw the horrific Abu Graib images:

> El origen de la utilización de los cuerpos desnudos forma parte del proceso preliminar de *Otras geologías*, cuando estaba realizando las investigaciones del proyecto salieron las imágenes de la prisión de Abu Graib, me impactó el tratamiento del cuerpo humano como un residuo. En las imágenes había unos cuerpos acumulados, formaban una montaña de cuerpos humanos, en concreto estoy pensando en una fotografía donde los cuerpos de unos prisioneros estaban agolpados en el suelo y tenían bolsas de basura en la cabeza.

> (The trigger to use nude bodies is part of the process preliminary to *Otras geologías*. While I was doing research for the project, the images from the Abu Graib prison came up on the media, and the treatment of the naked body as residue had a profound impact on me. The images showed bodies on top of each other, a mountain of bodies, and I am thinking about one particular image, where prisoners on the floor had their heads covered with trash bags.)

> (Ibarz 2009)

The year Jordan and Canogar made their trash art pieces public, 2005, was the year also when Fernando Botero exhibited his series on Abu Graib for the first time.[3] The obscenity of excess (excessive consumption, excessive trash, excessive violence), and of the disposability of human life has had a deep effect on the three artists, and has also informed the creative production of other well-known painters, sculptors, and media artists working with and on 'monumental' rubbish, among them Vic Muniz and HA Schult.

Vic Muniz's famous artistic intervention in Jardim Gramacho, Rio de Janeiro's biggest landfill before it closed down in 2012, also caters to the megasized nature of trash accumulation, in the style of Jordan and Canogar. And like Francisco de Pájaro's street art, Muniz's trash compositions equally acknowledge the powerful link that ties trash to poverty. It is not by chance that the 'true' manipulators and collagists of trash in Vic Muniz's monumental trash compositions are the former rag pickers and inhabitants of Jardim Gramacho. However, the emphasis on the trash–poverty continuum has very different aesthetic outcomes and political implications in the case of the Spanish artist and his Brazilian counterpart. De Pájaro's street art is deliberately unpolished and rough, with ugliness invariably overpowering any attempt at artistic embellishment. Muniz's 'landfill art', on the other hand, aims at beauty pure and lofty. The ugliness of the primary material (trash) needs to be transcended at any cost, and is required to render homage to old masterworks. Sure enough, Muniz skilfully directs his team of '*catadores*' (rag pickers in Portuguese), and teaches them how to construe replicas of famous paintings, such as Goya's *Saturn Devouring His Son*, Caravaggio's *Narcissus*, or David's *The Death of Marat*. These 'mimicking' collages first lie flat on the ground, and, upon its completion, are photographed from a high platform. The end result of this landfill project, known as *The Pictures of Junk* (2008) is 'a series of seven images, each produced in two different formats, with 50% of the sales of the larger prints benefitting the Garbage Pickers Association in Jardim Gramacho' (Hunt 2012).

Despite the evident connection of *Pictures of Junk* to poverty and even depiction of the victims of poverty and squalor (Muniz used a number of '*catadores*' as models for some of its trash collages), Muniz's landfill art is more closely related to the loftiness of Canogar's *Otras geologías* and of Jordan's *Intolerable Beauty* than to de Pájaro's down-to-earth '*El arte es basura*' urban trash creatures. 'Down-to-earth', in fact, is what does the trick, or rather, explains the fundamental difference of de Pájaro's approach to trash. More often than not his 'trash people' sit, lie, or crouch on the pavement, and passers-by are forced to lower their eyes in order to acknowledge their presence. De Pájaro does not have the privilege of vertical canvas to show off his art, nor are the viewers of his art allowed to enjoy beauty comfortably displayed at eye level. De Pájaro's art is at realist and immediate as it gets, because it favours 'real' horizontality over 'stylized' verticality. Poverty and trash happen on the floor, they do not hang from walls. In order to portray (and denounce) poverty, de Pájaro kneels down, and smells and touches trash, instead of climbing on a platform, and taking odourless and pristinely clean photographs from above, or even 'directing the installation from [a] studio and with the help of a laser pointer' (Hunt 2012). In this aspect, Canogar's approach to trash is very similar to that of Vic Muniz, since he also takes pictures of trash, and uses them to painstakingly construe a giant collage or composition.

Muniz, Jordan, and Canogar favour verticality over horizontality, distance over proximity, cleanliness over dirtiness, the odourless over the smelly, and ... durability over perishability, though in varying degrees: Canogar, for example, throws away his enormous trash collages once the exhibition is over. As the artist himself explains,

> Las fotografías son encoladas directamente en la pared y al acabar la exposición se arrancan de los tabiques y entran en el ciclo de la basura, quería que el proceso fuera éste, son obras efímeras con un carácter de autodestrucción programado desde su existencia.

(Photographs are directly plastered to museum walls, and, once the exhibition is over there are torn from the walls and enter the cycle of trash. I wanted the process to be like this, since [these photographic murals] are perishable goods, with self-destruction programmed into their existence.)

(Ibarz 2009)

Canogar thus is keen on restoring trash to trash, whereas Muniz seems more fixed on permanence and on art's survival and continuity. First of all, his trash installations are lifted from the landfill floor, via photography, and rescued from sharing the destiny of trash/us (which is to lie down, and rot); but more importantly, the permanence of his particular variant of trash art also prolongs the life of art, of high art in particular, and since his own work imitates and pays homage to the works of the great Masters.

Verticality, distance, and durability are in many ways the landmarks of high culture and its artefacts. And so are the monumental and the gigantic (Stewart), for to 'think big', and to 'do big' is the privilege of the rich and powerful. It comes as no surprise, then, that fairly affluent artists (Jordan, Canogar, Muniz) are the ones who embrace vertical, distant, permanent, and monumental, whereas artists of lesser means (de Pájaro) stick to horizontal, nearby, perishable, and small. Certainly, trash shares all these traits (almost all of them), with de Pájaro, and therefore one could argue that de Pájaro's art is much closer to trash, and an infinitely more accurate depiction and representation of trash than the art of Jordan, Muniz, and Canogar. Trash lies on the floor (is horizontal), is graspable and always around (certainly not distant, although we would like it to be), and it is perishable (or so we hope). But ... is trash monumental? Trash starts small (as small as the peel of an orange, or even smaller, like a cigarette butt, or a bus ticket distractedly made into a roll), but it grows fast, thanks to the powerful: humans. So if trash becomes huge and monumental it is because of us, because we, mighty colonizers of planet Earth, force it to expand, to move, and to accumulate. In an interview, Canogar not only speaks to the monumentality of trash, but also to its ability to dangerously turn against its creator, and to bury humankind under its pestilent weight. When asked why he uses photography to give expression to the seemingly unstoppable accumulation of waste, he responds:

El medio que he utilizado es el mural fotográfico, que desde mi punto de vista es diferente de la simple fotografía. Con estas imágenes quería llenar la mirada del público. El problema de la acumulación de residuos es un problema a gran escala, por ello era importante hacer las fotografías a gran escala. Al ser unas obras tan grandes, monumentales ... pretendía dar la sensación de que todos los residuos que aparecen representados en la imagen, se caían encima del espectador.

(The medium I decided to use is the photographic mural, which, from my point of view, is different to photography. My goal with these images was to fill the eyes of the public. The problem of residue accumulation is a problem of big scale, which is why it is important to take pictures that are equally oversized. My photographic murals are so big and monumental because I want to convey the impression that all the residues depicted in the image would tumble down and bury the spectator.)

(Ibarz 2009)

Monumentality and movement go hand in hand. Trash grows because it moves from hand to trash basket, from trash basket to dustbin and dumpster, from dumpster to garbage truck and from garbage truck or barge to transfer station, and, finally, to its ultimate destination point, a dump or a sanitary landfill. Artists look at trash, and make it the prime material of their work at all its different stages of growth and travel; some of them embrace trash when it is small, still discernible (with a face even), and has not yet travelled far (de Pájaro, 'My Dog Sighs', 'Filthy Luker', Laderman Ukeles, *Project TrashCam*); others meet trash when it has reached its final destination, is huge, and a landscape (Jordan, Canogar, Muniz); and some even stress the nomadic spirit of trash, its restless nature. Such is the case of German artist HA Schult, the creator of an imposing army of robot-like rubbish 'soldiers' that travels the world and takes famous tourist sites by assault.

Trash People. In twenty containers they roam around the world like refugees of the consumer society. The trash people are images of ourselves. We produce trash and we will become trash. Today's coca cola bottle is the Roman archeological found [*sic*] of tomorrow,

(Schult 1999)

says Schult about his creatures. He stresses their travelling nature, but he also adds a sombre accent to it. For 'Trash people' are not tourists but 'refugees' (refugees of our consumer society) who 'roam the word in twenty containers'. Once let loose, though, and ironically enough, they find themselves among tourists, having their picture taken on the Great Wall of China or in front of the Pyramid of Cheops. But, unlike tourists, who are always on the lookout for new attractions, the same 'trash people' come back, again and again. Trash never leaves us and always returns, no matter how far we wish to send trash away. And trash, as de Pájaro already vaticinated, 'is always the same'. In fact, Schult's trash sculptures are deliberately built to look alike, with identical body postures and marching in rigid unison.

The monumentality of Schult's trash automatons – the result of sheer quantity, imposing size, and expressionless uniformity – stands in stark contrast to the unassuming airs of de Pájaro's animated trash bags. In both cases the raw material is the same, waste, and the initial purpose similar, namely, to endow refuse with anthropomorphic features. But the ultimate goal is very different, and has to do with what aspect of trash (and of society) speaks to each artist. De Pájaro's basic purpose is to denounce poverty, and to establish a crass comparison between the two (interchangeable) entities consumerist capitalism so easily turns a blind eye on and discards: things, the poor. Schult's aim, on the contrary, is to raise awareness about waste as an environmental catastrophe of the vastest proportions, and, true enough, as a dire consequence of savage consumerism. In their acerbic critique of global capitalism, de Pájaro and Schult stand united. Still, in the eyes of the latter, trash is a monumental force that disengages itself powerfully from the floor; it does not loiter, but moves decidedly ahead. Schult, Jordan, and Canogar share the same apocalyptic vision of trash. Garbage, the consequence of our criminal irresponsibility, engulfs us, invades the planet, and is here to haunt us. Schult's 'trash people' are as close as they can get to zombies, and Schult's travelling installation certainly bears an uncanny resemblance to the zombie-film genre.

De Pájaro and Canogar, as well as Jordan and Schult, approach waste from different angles, but the works on trash by the four artists convey a similar mixture of fascination and horror. *Intolerable Beauty*, the title of Jordan's artistic project, is probably the most fitting description of Trashtopia, for its dystopic nature is certainly jarringly beautiful most of the time, and never entirely devoid of utopic elements. This is particularly true for de Pájaro's 'down-to-earth' art (both literally and figuratively), and its deeply compassionate approach to all things (and people) discarded. De Pájaro's art meets trash at the preliminary stage of its voyage and life cycle; still an infant so to speak, it is much easier to identify with it, and to feel compassion for it at that early phase than later on, when it has it has grown into monumental danger. But no matter how 'small' (and harmless) or how 'big' (and harmful), trash, in the works of de Pájaro, Jordan, Canogar, and Schult, showcases two essential traits: it travels, and it is always the same. '*La basura siempre es igual, hay basura por todas partes*', de Pájaro told me, and, sure enough, he found trash (the same kind) and poverty (the same kind) first in Barcelona, then in London, and lately in New York. Trash is everywhere. It spills out of bins and dumpsters, fills landfills to their capacity, piles up against the background of the Pyramids of Egypt, and crowds the Great Wall of China. Garbage is global, and Trashtopia, a monumental landscape that endlessly repeats itself across the globe.

Notes

1 See image gallery at: www.franciscodepajaro.com/#!art-is-trash-street-art/c2zv.
2 See image gallery at: www.danielcanogar.com/ficha.php?year=2005&proyecto=01_otrasgeologias&lang=es&foto=2.
3 See image gallery at: www.zonaeuropa.com/20050413_2.htm.

3 Dirty familiars

Colonial encounters in African cities

Stephanie Newell

> We need much deeper knowledge about the ways in which modernity has historically evolved in the cities of the non-Western world, what urban constellations and conflicts it has created there, and what such developments might mean today for city cultures at large.
>
> (Huyssen 2008: 2)

While travelling with William Hesketh Lever and others on a lengthy tour of inspection of Lever Brothers' numerous trading stations in West and Central Africa, in the mid-1920s, Thomas Malcolm Knox (1900–1980), secretary to Lever, adopted an interpretive framework that is both depressingly familiar today in anti-cosmopolitan discourses, and richly symptomatic of the economic relationships embedded in encounters between strangers in urban environments. The city of Lagos, Knox (9 January 1925: 72) noted, 'turns out to be a town of unspeakable squalor. It is no wonder that it is the nurse of disease. Filth everywhere.' For Knox, the source of filth was easy to identify, for '[e]verything reeks of dirty natives' (9 January 1925: 72). Yet this same city, he recognized, 'is the representative of a much higher state of civilization' than the 'squalid' African trading posts he recently visited in the hinterland of the Belgian Congo, for Lagos boasts European shops built to supply local consumers with household products manufactured in Europe, using raw materials exported from Africa's 'uncivilized' interior (Knox 9 January 1925: 72).

Several scholars (McClintock 1995; Burke 1996) have commented on the circular, self-serving nature of the connection between cleanliness and civilization in the writings of European travellers during the colonial era. Perceived and narrated through 'imperial eyes' (Pratt 1992), the figure of the 'dirty native' legitimized European cultural expansion into the most intimate corners of Africans' daily lives. In the eyes of imperial commentators, 'dirty natives' were far more dangerous than objects discarded by the wayside, or urban trash, and their ubiquitous presence in colonial cities caused colonial governments to enforce regimes of sanitation and urban racial segregation. Not coincidentally, these regimes also helped to transform imported luxury manufactures such as Lifebuoy Soap, Sunlight Soap, Lux, and Vim – all produced by Lever Brothers in the UK – into household necessities for urban consumers in global locations (see Figure 3.1; Burke 1996; Allman and Tashjian 2000).

In a book about globalization, garbage, and the contemporary city, it may appear paradoxical, if not perverse, to begin a chapter with a focus upon late nineteenth- and early twentieth-century intercultural encounters that do not fit current models of globalization and urbanization, in which high-speed communications and mass travel make possible the rapid repositioning of people and objects, consumers and commodities, discardable and recyclable materials, through international economic and cultural circuits. Indeed, as demonstrated by the recent media coverage of the transmission of Ebola within and beyond West African cities, a key feature of globalization is the potentially uncontrollable rapidity with which people are able to move from place to place.

Knox's negative responses to the strangeness of others in the 1920s, however, and the similar reactions of numerous other European travellers and traders in Africa in the late-nineteenth and early twentieth centuries, provide us with a historically situated, context-specific prologue to an unpalatable side of current discourses about globalization and urbanization. Such a prologue is embedded in the economic, political, and discursive power-relations that underpin contemporary global networks of trade and migration. Knox, and the other colonial white men who feature in this chapter, are not presented – or made present – as individual travellers in Africa who somehow merit biographical visibility above the Africans they describe. Rather, they feature as vectors for a distinctive, shared anti-cosmopolitan discourse that reaches back to the early days of empire and exploration, and continues to have global currency in contemporary public discourses about migration and mobility between global cities.

This chapter uses examples from imperial-era travellers' and traders' journals to suggest that an understanding of cross-cultural relationships from the past can help us contextualize the anti-cosmopolitan current that remains prevalent within contemporary debates about multiculturalism and migration, particularly in urban environments. In countless colonial-era travelogues and memoirs by British white men, a hermeneutic similar to that in Knox's diaries operates, leading to the same dead-end conclusions each time. A rhetoric of difference is mapped onto the body of others through a spectrum of dirt-related words. In one undated memoir by an anonymous trader (Anonymous n.d.b: 164) who worked for a Lever Brothers franchise in the early twentieth century, for example, 'the bushmen tribes' are described thus: 'Not only did their bodies give off a horrible smell, but their hair was tousled like dirty rope, and their skin a dull black. The bits of cloth around the loins were pregnant with filth.' In an earlier, similar, example from a memoir by the trader John Whitford (1877: 125–6, 160), one 'hideous-looking ju-ju man' is regarded as 'filthy' not because of his unwashed status but because he localizes and subverts imported items such as Western clothing and rum for his own cultural ends, wearing clothing in the 'wrong' way rather than as it was intended to be worn by the manufacturers.

Reiterated by numerous colonial travellers in Africa, these conclusions are not simply, or solely, a case of closed ethnic categories being pasted onto the

Lux Will Wash Locks.

[Photo by M. Frost, Biddenden.

Here is a lady of colour, hailing from Africa, preparing to wash her luxuriant locks in Lux, which, of course, is famous as a shampoo soap. The jolly subject of the photograph uses Lux regularly, and, judging by the crop shown around her head, it acts as a fine tonic for the hair.

Figure 3.1 Advertisement for 'Lux' in *Progress Magazine*, a Lever Brothers publication. Notice the hand-drawn insertion of the carton of Lux in the photograph (source: *Progress Magazine: The Magazine of Lever Brothers and Unilever Ltd*, July 1925: n.p. Reproduced with kind permission of Unilever from originals in Unilever Archives).

'other' by colonial selves located on the moral high ground in the decades before African independence. As the latter part of this chapter will argue, Knox and his contemporaries' discourse has a vibrant historicity that reverberates through the decades, changing with the times but permeating how the bodies of migrants and strangers are observed and produced by those with the power to tell stories and to be heard in present-day global contexts.

As Knox and his fellow travellers in the mid-1920s[1] moved southwards into Arab-Islamic Africa and beyond, a visceral hermeneutic increasingly dominated their descriptions of the inhabitants of the new urban environments they encountered. In the Moroccan city of Casablanca, Knox

> was disgusted with one of the main streets. On one side, bazaars: carpets and brass work hanging out over the street ... on the other side *Brasseries Majestie* and the like – dirt and filth – there is something most repulsive to me in these endless *Brasseries*, all dirty, crowded with dirty people drinking dirty looking drinks.
>
> (Knox 29 September 1924: 3)

This connection between the category of dirt and the consuming body of the 'other' starts in Europe for Knox: 'I was repelled by it all in Brussels', he notes, although, for no stated reason, 'it is ten times more repulsive in Casablanca' (29 September 1924: 3). From Casablanca onward, Knox and his fellows passed through urban environments where African and Arab traders from diverse religions and ethnicities interacted cross-regionally with traders and consumers from other towns and cities; where African women and men interacted with European men and women in intimate and domestic relationships, as well as in public urban spaces; and where European colonial officials undertook to rule in collaboration with local African elites, including chiefs, churchmen, educators, lawyers, imams, traders, local intelligentsias, and political activists. Each set of relationships brought its own cross-cultural efforts to interpret and understand the other.

What revolts Knox in Casablanca is the entire public *habitus*: the busy local cafés and the messy protuberance of local commodities displayed for sale in the shops. What revolts him, in short, is the presence of the foreign body as a consuming entity that participates in a cash economy but desires merchandise that is completely alien to his own trade interests, illustrated in Figures 3.2 and 3.3, which show the goods on display in typical Niger Company stores in the early twentieth century. The powerful physical feeling of revulsion that he experiences marks the moment at which Knox recognizes the other's humanity as a consuming subject – eating, drinking, socializing, purchasing goods – and instantaneously disavows the other's tastes as unpalatable to himself. This argument differs somewhat from theorizations of dirt as that which society expels, excretes, or treats as abject or excessive (Bataille 1985; Douglas 2002 [1966]; Kristeva 1982; Smith 2007). In these accounts, dirt figures as a category that mediates between the margins and the mainstream, facilitating the expulsion of particular types of matter from social visibility.

Figure 3.2 'Photograph of the interior of a trading store, Nigeria, and the goods available, 1920s–1930s' (source: UAC/1/11/9/3/27. Reproduced with kind permission of Unilever from originals in Unilever Archives).

Figure 3.3 'Photograph of the interior of a trading store, Nigeria, 1920s–1930s' (source: UAC/1/11/9/3/69. Reproduced with kind permission of Unilever from originals in Unilever Archives).

The category of dirt cannot be divorced from the judgement of people's dirty *habits* (in both senses of habit, as clothing plus lifestyle choices). What appears to the onlooker's eyes to be the dirtiness of others generates clear moral and political judgements about their behaviour and lifestyles. In spite of his protestations of loyalty to judgements based upon the observation of 'empirical phenomena' (18 February 1925: 100) in Africa, Knox's disgust and repulsion are not focused upon unclean streets or unwashed bodies, but upon unrecognizable objects. In Jebba, Nigeria, for example, he describes how 'We stopped at various native stalls and examined their wares – capsicum (pepper of a particularly strong variety), chop of various sorts, extraordinary and repulsive stuff all of it' (9 January 1925: 75); at the market in Zaria, Northern Nigeria, he finds that, '[t]he meat presents the most disgusting appearance. It is covered with flies and vermin and even were these absent seemed to consist mainly of the least savoury looking parts of animals' (17 January 1925: 81). Also in Zaria market, he finds that, 'The knick-knack stalls were the most curious of all. Little bits of stick, a few knobs of ginger, little bits of stone, a tooth pick or two, all apparently things of little or no use' (17 January 1925: 81–2). Locally manufactured African products are regarded as inferior to their imported counterparts, their very presence marking a lack of civilization among native consumers (see Burke 1996; McClintock 1995).

All the way from Casablanca to the interior of Congo, the marked preference of local people for locally produced commodities – unrecognizable to the traveller – above imported commodities purchased from the European companies operating in the region filled commentators with revulsion and rendered the local a nauseating other, who resisted assimilation into the global economy represented by the European trader. These visceral responses have little to do with dirt as an empirical substance, and more to do with white traders' subjective reactions to local consumption practices. As one anonymous memoirist recalled of his time as a trader in Sapele (Nigeria), while European company workers exchanged Trade Gin with Africans for rubber, '[s]avage and unclean hunters, almost naked', arrived on the scene with their own trade-goods, carrying, amongst a range of unrecognizable objects, bush-meat and 'monkeys and gray parrots for sale' (Anonymous n.d.a: 5). An earlier trader on the River Niger, John Whitford, was similarly revolted by the unfamiliar appearance of local people and goods, especially the women: for Whitford in the 1870s, villages were 'filthy' because they contained 'hideously ugly' women, whose ugliness stemmed from their 'strong limbs developed by hard work, which should pertain to a man only' (1967 [1877]: 142).

The category of dirt signifies disorder, inefficiency, and the unrecognizable, in these traders' accounts of the continent. In this, their interpretive framework conforms to the anthropologist Mary Douglas's resonant assertion that dirt marks the limits of a society's understanding of itself and signifies people's need to withdraw from any habitus that is perceived to be dirty, and, in reaction, to reassert their own interpretive boundaries (2002 [1966]). For Douglas, as Richard Fardon notes, 'ideas of impurity and danger hold members of a society to account to one another, and they do so with a character and intensity that stems from and rebounds back upon that particular form of society' (Fardon forthcoming: 7–8).

Disorder is not the stopping point for Knox and his fellows, however: the out-of-place-ness they attribute to 'dirty natives' is further translated into the visceral category of disgust. In so doing, the materiality of people and things is firmly resituated in a moral realm: visual observations about the stranger's consumption habits are processed by Knox and his peers into domestically meaningful opinions (and projections) about the stranger's *habitus*. Interestingly, invisible 'sights' are also projected into this set of opinions. The army officer, Captain Alan Field, author of a bestselling book of advice for first-time travellers to West Africa, summed up European suspicions, asserting (1913: 49, 144) that 'Africa as a country [*sic*] tries to conceal the vileness of current flowing through it'. From the visual perception of other people's consumption practices – or, as in the above citation, the paradoxical 'observation' of concealed social phenomena such as cannibalism – arises the visceral representation of Africans as disgusting, and thus as morally and ontologically inferior. Dirt is the mediating category for this intercultural encounter.

Commentators such as Knox and Field were writing in an era that followed the exposure of dreadful slum conditions in British industrial cities in the 1890s, and Knox's journal occasionally compares the squalor of native quarters in African towns with the slums in British industrial cities (4 November 1924: 42, 18 February 1925: 99). Such a conflation of British dirt and dirt in colonial contexts had, however, become dangerously unsettling to the British imperial mission by the time Knox wrote his journals. Ethnographic accounts of dirt in Africa had been overtly referenced in sociological accounts of the dire conditions of British working classes in industrial cities, epitomized by William Booth's *In Darkest London and the Way Out* (1890), published in the wake of Henry Morton Stanley's bestselling and much-serialized *In Darkest Africa: Or the Quest, Rescue, and Retreat of Emin Pasha, Governor of Equatoria* (1890). Similarly, Margaret Harkness's (1889) account of London slums, originally entitled *Captain Lobe: A Story of the Salvation Army* was changed immediately, upon the publication of Stanley's volumes, to *In Darkest London*. These social reformers' conflation of 'white' dirt with African dirt added political weight to their campaigns to halt so-called primitive conditions in British industrial cities. Their use of the metaphor of 'darkness' for domestic cities created a shock value for audiences familiar with the descriptions of African cultural darkness to be found in Stanley's celebrated journals and the popular literature they inspired.

Dirt was a key ingredient in the making of imperial identities and in the marketing of imperial products to global consumers in the colonial era. Out of it grew new global markets to the extent that, in one advert at least, soap as a commodity replaced the Victorian moral principle that 'cleanliness is godliness'. A famous advert from the Pears Soap Company in 1890 starkly reminded consumers, via a misquotation from Justus von Leibig, that '[t]he consumption of soap is a measure of the wealth, civilisation, health, and purity of the people' (see Figure 3.4).[2]

Figure 3.4 'Advertisement for Pears Soap' (source: *The Graphic*, 30 April 1890: 36, author's collection).

Significantly, this advert is printed on the back cover of a special 'Stanley Edition' of *The Graphic* celebrating the recent Emin Pasha Relief Expedition led by Stanley from 1886 to 1889. Through numerous vivid drawings of Stanley's expedition from the east coast of Africa into the interior of the continent, the special issue illustrates the contrasts between the moral authority, bravery, and leadership of British men, and Africans' lack of control, illustrated not least through their lack of clothing (Figure 3.5). Appearing in the wake of these images and reports, the Pears Soap advertisement on the back cover is a tangible by-product of the 'dirty native' ideology.

As a marker of ontological differences between the self and the other, the category of dirt usefully confirms the beholder's sense of superiority over the rejected body. Colonial commentators had to maintain the boundary between self and other, and not allow the category of dirt to shift out of place: if white working-class dirt had coincided with African dirt, the ideological and moral foundations of the so-called civilizing mission would have collapsed. Dirt 'sticks' precisely because it is presented as descriptor of visible features: revulsion and disgust are experienced by the beholder as natural, biological, sense-based responses to observable phenomena, rather than as personal imaginative failures in the onlooker's effort to comprehend new environments.

Knox, Field, and Whitford were not alone in their feelings of revulsion toward the consumption practices of strangers. For these cross-cultural commentators, however, whose livelihoods depended upon the expansion of European trade and the development of local markets in Africa, the perception of dirt was riven with ambivalence. As an industrialist seeking to expand global demand for the household products manufactured by Lever Brothers, as well as to secure an efficient workforce of African wage-labourers for the extraction of raw materials in the colonies, Knox continuously had to resist his visceral revulsion and attempt to redomesticate the other body. As he points out in a later report to the Chairman and Board of the United African Company, 'the successful European trader is he who can mould the native taste' (Knox 1929), not he who passes by, holding his nose in disgust (Knox 1924–1927: 4). For the white trader in Africa, large profits were to be made from successfully anticipating 'native taste' for particular imported items in exchange for local produce such as rubber, palm nuts, and palm oil. As one anonymous trader's boss stated starkly in a word of advice, 'There is no doubt if you can get the patterns on the cottons printed in colours the Bushman likes, they'll bring plenty of trade' (Anonymous n.d.a: 72).

Traders' feelings of alienation and revulsion can be understood as their physical internalization of other people's resistance to capitalist expansion. William Hulme Lever noted in a letter to his father from an earlier expedition, '[t]he Pagans are very disinclined to work, their wants being so few', rendering 'the possibility of trading with them … very limited' (5 August 1921: 8–9). The aim of trade was to transform the natives from 'idle loafers into real workers' with a weekly wage to spend on consumer goods (Knox 29 October 1924: 34). In other words, the trader's feelings of revulsion signify his failure to 'mould the native

FOREST DWARFS EATING SNAKES

The dwarfs, who are about four feet in height, with bodies of a rich brown colour, entirely covered with a soft short down, are unable to live in the sun, and remain almost entirely in the forest, under the shade of the great trees. They live almost entirely on roots and vegetables, but occasionally indulge in snake

Figure 3.5 'Forest Dwarfs Eating Snakes' (source: *The Graphic*, 30 April 1890: 10, author's collection).

taste'. Thoroughly affected by economic factors such as the expansion of industrial European markets into colonial households, and the creation of cash economies through the employment of African wage-labourers, those who are repelled by the 'native' experience the full visceral affects of local consumers' resistance to the expansion of industrial capital into their own economies. By contrast, local consumers who 'became more intimate with the whiteman, and made money in trade', also became more physically palatable to the European trader by the fact that 'they went in for fancy [imported] cloths as well' as local materials (Anonymous n.d.a: 164). Such a person could never resemble the white man, for 'even when they purchased cheap perfumery ... the other "scent" was the stronger' (Anonymous n.d.a: 164).

Knox's travels with his employer leave a great deal unsaid about the use of forced labour in Lever Brothers' Congo concessions, meticulously catalogued in Jules Marchal's *Lord Leverhulme's Ghosts: Colonial Exploitation in the Congo* (2008). As with many imperial traders before him, Knox's vision is inward-looking rather than externally focused as he continuously attempts to overcome the presence of the resistant 'native'. In an effort to recognize the strangeness of Africa in terms that are familiar, he uses similes to reach across the vast geographical space separating 'home' from West and Central Africa. As a Scotsman, he finds that the journey from Boma to Matadi in the Congo is 'similar to that up Loch Fyne', while up-river, 'I was really reminded a bit of Killiecrankie' (20 October 1924: 11, 12). The further up-river he travels, however, the greater the strain on his similes: further on into the Belgian Congo, he finds, 'It is Inverness to the Kyle of Lochalsh *only more so*' (20 October 1924: 16; emphasis added). Commenting on the territory as a whole, he writes, 'for real beauty and grandeur I am bound to say that I prefer the Rhine gorge and, in some ways, Loch Fynne; *but it is very hard to make comparisons*' (17 October 1924: 20; emphasis added).

These similes mark the onlooker's efforts to reach across an untranslatable divide in order to retain contact with potential markets and consumers. Knox's similes are, however, used almost exclusively to describe landscapes and vistas. The towns and cities he and his fellow travellers pass through are filled with people, and as a consequence of the human element, '[t]his tropical beauty was full of danger and disease' (Anonymous n.d.a: 79). Landscapes, buildings, and objects can be clean, but only *people* (and the things people do and sell and touch) can be dirty for the imperial traveller. People and communities attract dirt-related words: they are ugly, revolting, disgusting, diseased, and filthy. Such words mark the end point of a process of cross-cultural interpretation that moves via similes and likenesses towards a failure of recognition of the stranger as a fellow human being. This discourse is familiar in current reactions to the transmission of the Ebola virus, as examined in more detail below.

Not all imperial travellers and traders adopted this anti-cosmopolitan discourse. Many white men developed intimate and loving relationships with local people, and celebrated the beauty of life in mud huts with clay floors, local 'wives', and home grown foodstuffs (Anonymous n.d.a: 107). For one anonymous trader, life is rendered exquisite in Nigeria by the presence of his beloved companion, known to him as 'Ting a Ling' (Anonymous n.d.b). For another long-term resident in West Africa, John Moray Stuart-Young, the cleanliness of the bustling market town of Onitsha was incomparable to the dirt of the Manchester slums in which he was born (Newell 2006). Such men were seen to have 'gone native' by anti-cosmopolitan commentators, their consumption habits dangerously unsettling the imperial right to rule.

Dirt as interpretive failure

If dirt functions as a category of urban understanding, and as an interpretive tool in multicultural environments, it is deployed viscerally rather than deliberatively by the people who use it as a mode of comparison between the self and others, between neighbours and strangers. The category of dirt puts matter firmly into place, fixes it in an interpretive hierarchy that relates to the sight and behaviour of people, and draws what is hidden into view. In short, dirt *matters*: it is visible, materialized physically and viscerally to the beholder through sight, smell, touch, hearing, and taste, and it is based – for the beholder at least – upon incontrovertible physiological reactions such as nausea and disgust that spawn value-judgements about the behaviour of others.

In his analysis of European 'orientalist' modes of perception in the nineteenth century, Edward Said describes the manner in which the repetition of particular analogies about the Arab-Islamic world '*create* not only knowledge but the very reality they appear to describe. In time, such knowledge and reality produce a tradition' (1978: 94; emphasis in original). For Said, the recurrent use of particular labels and terms of comparison in a dominant discourse will, over time, take a hold of the existential complexity of the person described and replace it with an essence, a condensed and repetitious metonym that stands in for – and takes over from – the whole. For Said, after Foucault, representations and repetitions serve to define realities: in this way, the dominant group's power is exercised through the production of knowledge about 'the Orient'.

Said regards the production of knowledge as a sign of discursive hegemony, but dirt as a category for knowledge production contains within itself a complex process of aspiration and failure. Dirt marks an effort to understand and cross into the world of the other, a realization of the failure of empathy (or humanist understanding), and a retreat into the sense perceptions of revulsion and disgust.

Numerous negative cross-cultural encounters in the twentieth century are mediated in this way. In his classic study of colonial racism, *Black Skin, White Masks (Peau noire, masques blanc)*, for example, Frantz Fanon (1967 [1952]: 109–40) reports on the traumatic moment when, while walking anonymously down a post-war Parisian street, his complex humanity is suddenly reduced to a state of epidermal otherness. His identity is fixed permanently in place, skin-deep, by a white person who reacts to the sight of him with the words, '*Sale nègre!*'. The racial descriptor, *nègre*, is inextricable from the adjective, *sale*. There are at least 22 different translations of the French word *sale* (also *la saleté*), all with

negative connotations that resonate, cumulatively, through its most basic translation as 'black', including dirty, smutty, trashy, grubby, foul, messy, oozing,
depraved, obscene, greasy, nasty, unclean, disgusting, and defiled.

The fear of defilement experienced by Fanon's white observer in the 1950s is
reiterated many times over in racist speech. In 1958 in Congo, for example, the
doctor Jacques Courtejoie recollected,

> two months after independence I went to dinner at the home of a white
> regional administrator. He came home late, because he'd been to a political
> meeting of the Abako. When he got home, his wife said: 'I certainly hope
> you didn't shake Kasavubu's hand!'. I can still hear the way she said that.
> Even by that time, people still thought Africans were dirty!
>
> (Van Reybrouck 2014: 290)

These sentiments persist in current Western popular reactions to Ebola, a virus
that can only be transmitted by direct contact with the body fluids of an infected
person: thus in October 2014, a half-Sierra Leonean boy was excluded from a
primary school in Stockport, Cheshire, after a Facebook campaign against him
by parents of other children at the school (Robinson 2014).

A powerful parallel to this interpretive process can be found in the experiences of its targets. In his exposé of the physical and psychological affects of
racism, Fanon (1967 [1952]: 109–40) describes how racist representations stick
to their objects – just as 'dirt sticks' – contributing to and partly creating the
realities and ideas by which people live in metropolitan and colonial cities.
Sapping the subject's sense of self with powerful negative representations, the
observer fixes the other, influencing and shaping the identity of the target *on the
inside*. When he is hauled out of urban anonymity into the visual field of the
white passer-by on the streets of Paris, Fanon experiences physical nausea, or
self-revulsion, at the way his skin is recast by the gaze of the commentator.
These intimate encounters live within the psyche in the form of what Stuart Hall
terms an 'enigma', a 'tense and tortured dialogue', and cannot simply be
expunged from a culture or an encounter (Hall 1994: 400).

Obvious, if extreme, other examples of the conflation of dirt and cultural
otherness include the labelling of Jewish people as vermin by Nazis and European anti-Semites in the 1930s and 1940s, and the Rwandan genocide of 1994
that was initiated by a media campaign on *Radio Télévision Libres des Milles
Collines* (RTLMC) to 'exterminate/crush the cockroaches' (Prunier 1998). In
Rwanda, the media made use of the word 'cockroach' to relabel the Tutsi people
and their sympathizers as vermin; as with Nazi ideology in Europe, the analogy
with dirt became mass-murderous in the process. A similar discursive manoeuvre occurred in February 2014 when President Yahya Jammeh of the Gambia
described homosexuals as 'vermin' who should be tackled like malarial mosqui-
toes. Uganda's long-standing President, Yoweri Museveni, also explained his
introduction of new, toughened legislation against homosexuality using associations learned directly from the '*sale nègre*' discourse of dirt and contamination,

describing gay people as 'disgusting' and 'abnormal' (Landau *et al.* 2014). Within this logic, homosexual men (noun)=vermin (analogy)=filthy (judgement)=to be eliminated (social and legal response).

In each of the above expressions of revulsion from the 1950s to the present, a similar interpretive process to that used by Knox in the 1920s is employed, whereby culture (public opinion, arts, and media) imbues politics (government and the law) with perspectives that are predicated on the category of dirt, and where the category of dirt is rendered sensual, visceral, and material through reference to vermin, infestation, contamination, filth, and disgust. Discursively, as Foucault and Said argue, one can trace a pathway from mediated perceptions and opinions – newspapers, radio broadcasts, political speeches – to political and social outcomes at street level, and back to the media once again, reinforced by the truth-value of repetition.

The feature of this contemporary discourse that connects it to colonial-era discourse is that the expression of disgust – if not hate – is regarded by those who experience it as natural and instinctive rather than as ideological, precisely because it involves a set of visceral reactions to the habits and *habitus* of strangers. Thus, on reaching the town of N'Kunda in Congo on October 1924, Knox found that

> the people were thoroughly repulsive, especially the women. The people were all cicatrised in a way which does not leave smooth pictures on the skin but which, as it were, makes patterns of ulcers … on all parts of the body. It makes the people look repulsive
>
> (21 October 1924: 23)

Invoking ulcers – painful pus-filled sores – rather than culturally specific beautification practices on the female body, Knox connected medical infection with the aesthetic category of ugliness in order to legitimize his own feelings of revulsion.

Knox's dystopic vision of an Africa with 'children romping heedless of the endless flies and vermin' (29 September 1924: 3), where 'thoroughly repulsive' people (21 October 1924: 23) and 'degenerates' (23 November 1924: 55) live in 'dirt, grit, dust' (9 January 1925: 72), where 'deformed men and crippled children' intermingle with 'people wandering about suffering obviously from loathsome and unspeakable diseases' (21 January 1925: 85) is a century-old iteration of reactionary Western responses to African urban environments during periods of famine or disease. In the recent international media coverage of the spread of Ebola in Liberia, *The Telegraph* depicted New Kru Town in an apocalyptic language that is reminiscent of Knox's tone: 'sewage runs openly through its maze of corrugated shacks, and in Liberia's wet season – at its height right now – tropical torrents turn it into one vast, warm, moist, breeding pool for germs' (Freeman 2014). Crossing the physical and the cultural, the potency of dirt as a category for interpretation is its capacity to express taste (the judgement of beauty, ugliness, and sexual preference) in the form of 'gut reactions' to the proximity of the dirty body (nausea and revulsion), making the response of the beholder appear to be 'natural'.

Coda

Through dirt, this chapter has attempted to historically contextualize the phenomenon of anti-cosmopolitanism in postcolonial African cities, and to understand the ways in which contemporary urban relationships resonate with past ideologies in the form of politically charged reiterations of prejudiced discourses. The category of dirt is a particularly useful tool to access social prejudices in global cities, not least because it influences the terminology through which people and the media continue to interpret one another today. In October 2014, for example, among the comments on the UK *Daily Mail*'s website after an item on the spread of Ebola, one can find the following opinions about those in the West who allow African mobility to Britain: 'The liberal will soon spread this disease to all corners of the earth'; 'Please tell me all incoming flights are banned'; 'I shudder to think the consequences of people from these areas being allowed in and out of Britain' (Robinson 2014). Readers of the supposedly highbrow UK *Independent*, commenting on the exclusion of a West African boy from primary school, did not shy away from expressing similar sentiments: 'I hate to say it but they are similar to lepers. Touch them at the wrong time and you have it and die with the rest. So bring them here and risk possibly thousands of Deaths'; 'We should be closing our boarders [*sic*] to anyone travelling from West Africa' (Eleftheriou-Smith 2014).

In spite of the orientation of this chapter, however, it must be remembered that dirt as a category for the interpretation of otherness does not originate in European colonialism, nor do colonial encounters provide the exclusive source for current interpretations of others and otherness in African urban contexts. A multitude of words can be found in African languages to describe the dirt and dirtiness of others, dating back long before the colonial encounter. In the nineteenth century, Ndebele people used 'the Shona word *tsvina* (dirt) to describe their antagonists as *chiTsvina*, "dirty people"' (Burke 1996: 25–6). Similarly, evidence for the survival of precolonial concepts about dirt can be found in local ideas about 'ritual impurity' amongst Sotho and Tswana speakers (Brown and Beinart 2013).

What is missing from Knox's, Field's, Whitford's, and my own account above, and what remains largely absent from studies of hygiene and the history of public health in sub-Saharan Africa, are African perspectives on – and changing African historical understandings of – categories signifying dirt. Given the starting point of this chapter in the travel writings of imperial British men, and given the predominance of English language media in postcolonial Africa, dirt-related terms drawn from the English language have provided the primary source for this essay rather than African language terms and 'African local knowledge' (Brown and Beinart 2013). When African languages and creoles are added to the melting pot of colonial and postcolonial urban interactions, a proliferation of additional connotations and concepts arise, sometimes providing respite from (post)colonial discourses of hatred (Newell 2006; Epprecht 1998).

Similarly, when detached from the interpretation of others' ethnicity or sexuality, dirt-related terms in Africa are not always negative or extreme as in the cases of ethnic and sexual chauvinism described above. In some contexts, dirt-related words may be used to express positive evaluations of others, albeit in the form of jokes and proverbs that comment on a person's wealth (see Maranga-Musonye 2014). At a practical level, ordinary people's responses to globalization include frugality and the recycling of so-called trash (Coppoolse, Chapter 10 in this volume). The artistic transformation of 'garbage' and 'dirt' into beautiful, symbolic, or useful objects is also an area for further research (Born *et al.* 2012; Whiteley 2011). In several circumstances, therefore, visible signifiers of dirt may carry positive meanings.

In seeking to explore the opinions and discourses that fuel conflicted social relations in African cities, this chapter has sought to examine the flip-side of debates about global cosmopolitanism. The emphasis on negative and violent social interactions was not intended to suggest that African cities are any more (or less) culturally antagonistic than other cities globally. Contemporary African cities are characterized by urban dwellers' innovations in environments marked by extreme poverty and scarcity, as well as by multicultural inward migration from other regions. Urban subjects struggle to survive under increasingly severe constraints, while continuing to produce indigenous responses to the flows of local and international commodities and resources in their cities (Simone 2004, 2005). The provisional qualities of African cities and their instability are, in some scholars' views, the very features that allow for the emergence of creative responses on the streets (Barber 1987; de Boeck and Plissart 2004).

This chapter has sought to emphasize that globalization has a history in Africa that is refracted through the prism of diverse encounters, including the colonial encounter, that contributed to the continent's urban modernity. These urban encounters and identities – relationships with others, as well as the implementation of environmental and public health policies, and anti-racism initiatives – may be understood differently if they are filtered through concepts relating to dirt in its local and global manifestations, rather than concepts relating to hygiene and cleanliness.

The 'dirtying' of particular populations remains common in numerous global locations, from the fear of the spread of Ebola from the megacities of Freetown and Monrovia to Lagos, and these representations can become devastating under extreme political and economic conditions. The highly charged, politically productive discourses of racism, ethnocentrism, and homophobia stem from a history of (re)iterations of cultural difference through the supposedly empirical category of dirt.

In attempting to render the category of dirt productive, however, significant methodological and intellectual challenges arise relating to 'how to reconcile the universal and the particular in the practice of cultural criticism without lapsing either into empirical particularism or abstract universalism' (Huyssen 2008: 4).

Such challenges characterize the study of contemporary global urban cultures generally, but are of especial significance in cultural histories of postcolonial cities, where the temptation to source the (postcolonial) present in the (colonial) past through direct, connective comparisons risks reducing the former to the latter, and minimizing what Ash Amin (2012), AbdulMaliq Simone (2004), Arjun Appadurai (1996), and numerous other scholars of global cities highlight: if there is anything 'essential' about the global city, they insist, it is the unpredictable, productive potential of diverse 'urban imaginaries' in global contexts (Amin 2012: 68).

Furthermore, the very discourses analysed in this chapter contain within themselves a host of interpretive (or imaginative) failures and, crucially, an embodied, sense-perceptive consciousness of that failure. If regarded as a failed interpretive category rather than as a biological substance or a 'natural' response, our understanding of how dirt operates discursively can perhaps help us to comprehend, and confront, essentialist ideologies in diverse global settings. In other words, essentializing categories such as dirt are, paradoxically, filled with a vital historicity and cultural specificity that help to make visible the very production of universalist categories in response to particular urban encounters.

Acknowledgements

This chapter is part of an ERC research project, 'The Cultural Politics of Dirt in Africa, 1880–present' (ERC AdG 323343). I am grateful for the data provided by the researchers on this project: in Nairobi, Ann Kirori, Job Mwaura, and Rebeccah Onwong'a; in Lagos, Jane Nebe, Olutoyosi Tokun, and John Uwa. This document reflects only the author's views: the European Union is not liable for any use that may be made of the information contained therein.

I am indebted to Diane Backhouse at Unilever Archives & Records for allowing me access to the UAC archives at Port Sunlight over a period of several weeks, and for her kindness in bringing my attention to a wide array of materials.

Notes

1 At least seven other men travelled with Knox and Lord Leverhulme on this expedition, including William Hulme Lever (Lord Leverhulme's son and heir), Ernest Hyslop Bell (Chairman of the Niger Company), D'Arcy Cooper (Chairman of Niger Company), and Jonathan Simpson (architect and co-creator with Lord Leverhulme of the model town of Port Sunlight near Liverpool).
2 Leibig's actual words were:

> The quantity of soap consumed by a nation would be no inaccurate measure whereby to estimate its wealth and civilisation.... This consumption does not subserve sensual gratification, nor depend upon fashion, but upon the feeling of the beauty, comfort, and welfare, attendant upon cleanliness; and a regard to this feeling is coincident with wealth and civilisation ... [A] want of cleanliness is equivalent to insupportable misery and misfortune
>
> (1843: 18)

4 Waste not, want not

Garbage and the philosopher of the dump (*Waste Land* and *Estamira*)

Geoffrey Kantaris

The form of commodity is abstract and abstractness governs its whole orbit.

(Sohn-Rethel 1977: 19)

All of creation is abstract.

(*Estamira*, in Prado 2004)

Garbage, ethics, and the commodity form

In his analysis of the capitalist mode of production, Karl Marx gave pride of place to what he termed, with a little irony, the 'metaphysical subtleties and theological niceties' that abound in the commodity form (1976: 163). He endowed the commodity form, this bastard offspring of the coupling of dead capital and living labour, with a strangely animist half-life, for when a raw material such as wood, 'an ordinary sensuous thing', is transformed into a manufactured object such as a table,

> it not only stands with its feet on the ground, but, in relation to all other commodities, it stands on its head, and evolves out of its wooden brain grotesque ideas, far more wonderful than if it were to begin dancing of its own free will

(Marx 1976: 163–4)

Once set loose in the marketplace, these promiscuous dancing commodities, 'ready to exchange not only soul, but body, with each and every other commodity, be it more repulsive than Maritornes herself' (Marx 1976: 179), appear to take on a life of their own, independent of the human labour that originally animated them. For on the economic stage, Marx says, 'persons exist for one another merely as representatives and hence owners, of commodities [...]; it is as the bearers [*Träger*] of these economic relations that they come into contact with each other' (1976: 178–9). Social interaction is thus delegated to the relations between commodities, and the more lively becomes the movement of the commodities, the more human actions are reduced to those of automata, Golems mindlessly driven by commodity exchange, and the more we witness 'the conversion of things into persons and the conversion of persons into things [*Personifizierung der Sachen und Versachlichung der Personen*]' (Marx 1976: 209).

But what of the afterlife of these oddly animate craftings of sensuous matter? What happens when the commodities, as it were, stop dancing, and fall out of the spheres of both exchange value and use value? Of course, Marx's dancing tables were already presages of such an afterlife, since the analogy referred to the 'turning tables' used in séances during the spiritualist craze that spread through German upper-class society in the 1850s (Brookhenkel 2009). And it was this line of thinking – the mystical and spiritual investment in commodity production and exchange at the heart of bourgeois society – that to some extent determined Marx's application of the derogatory, primitivist vocabulary of 'fetishism' to the commodity form in *Capital*. Yet, other than the waste and devastation produced by capitalist crisis, Marx himself had little to say about the actual death (or spectral afterlife) of commodities, or about the places designated as their graveyards: the rubbish dumps or garbage heaps where commodities are sent once they are broken, or once their exchange value, even as raw material, falls below the perceived value of a new replacement.

Other thinkers in the Marxian tradition have partially explored this theoretical gap, albeit mostly in allegorical terms. Famously, Walter Benjamin developed a materialist aesthetics of the ragpicker (*chiffonnier*) out of Baudelaire's own fascination with the figure (Benjamin 2006: 52–4), recovering these members of the *Lumpenproletariat* from the historical dustbin to which Marx had confined them in the *Eighteenth Brumaire*, where they were lumped together with vagabonds, jailbirds, swindlers, *lazzaroni*, pickpockets, organ-grinders, tinkers, beggars, and other such reactionary layabouts (Marx 1975: 75). And just before his death, in 1940, Benjamin gives a Messianic force to 'the pile of debris' that the appalled Angel of History sees growing skyward as he rides the shockwave of that storm called capitalist 'progress' (1992: 249). The Angel would like to 'stay, awaken the dead, and make whole what has been smashed', if it were not for the fact that his wings are hopelessly caught in the storm's ferocious gale. All he can do is contemplate, with a melancholy gaze, the growing pile of 'wreckage upon wreckage [hurled] in front of his feet'. Benjamin's recuperation of refuse and of its collectors was also predicated on the surrealist penchant for cracking open the homogenous, empty time of the present with objects found in flea markets, ready-mades, or the exploration of the abandoned spaces of the city, and belongs to a tradition that assigns a subversive/redemptive quality to the bric-a-brac left behind by the crisis-ridden dreadnought of (urban) capitalist development and expansion.

Of course, waste and garbage, or trash, are not quite the same thing: the affective and ethical attributes attached to these words are of different conceptual orders. Trash is what has been trashed, ruined, or refused and needs to be removed, rendered invisible, as quickly as possible. There is little of an ethical dimension to refuse; rather, the act of refusal is the ground zero of ethics, the black hole into which ethics is swallowed, and to which Marx himself seemed blind, as suggested above. The act of designating other people as trash (whether it be the

'Lumpen' or the 'disposables' of Latin America's megacities) is better understood as pre-ethical, as something operating at the level of affect, or in an older vocabulary, libido, than at the level of an ethical regime. But waste is a whole different story. As the proverb in my title suggests, 'waste not, want not', the concept of waste, of a scarce resource that is irresponsibly deployed, is an ethical category par excellence, erecting the entire edifice of morality, from the virtues of thrift to religious injunction, passing through Marxist messianism, our Angel of the garbage tip, and ending perhaps in the threat of an ecological Apocalypse.

In societies based on commodity exchange, garbage is intimately related to the commodity form, being both its inevitable corollary and its antithesis as a mystified and abstract condensation of social relationships. To use another Benjaminian metaphor, we might say that *garbage is the commodity stripped of its 'aura'*. It is a thoroughly defetishized object that has fallen out of the realms of desire, exchange, and use, and has thus, in some sense, fallen outside of the realm of History, if we understand History as the product of a dialectic that has its origins in the division between intellectual and manual labour. For Marx, as is well known, the motor of history is class struggle, but the division of society into social classes is nothing other than the division between these different modes of labour – labour of the hand and labour of the head. Such a division is, however, only possible in a society where intellectual work can be exchanged for (the products of) manual work, and hence the division presupposes, and in large measure can be said to arise out of, the abstractions produced in and through the exchange of commodities. (This is the central insight of Alfred Sohn-Rethel's work, which will be discussed further below.) To say that garbage is a commodity that has fallen outside of the realm of History in the Marxist sense is not to deny the archaeological historicity of garbage, its role as a spectral record or remainder, nor the fact that renewed labour (such as that of the ragpickers) can reinsert garbage into the commodity cycle of exchange and use. But it is precisely as this spectral, indeterminate object lying at the ground zero of ethics, outside of the dialectic of history, that garbage can have a revelatory function, for Baudelaire, for Benjamin, for the Angel of History and, as we shall see, for the ragpickers themselves, even as the act of refusing refuse conforms to the logic of disavowal that supports the entire realm of commodity fetishism.

Extraordinary (global) garbage[1]

> What I really want to do is to be able to change the lives of a group of people with the same materials that they deal with every day.
>
> (Muniz, in Walker 2010: 0:06:20)

> For poetry makes nothing happen.
>
> (W.H. Auden, 'In Memory of W. B. Yeats')

The Brazilian artist and photographer Vik Muniz has become something of a celebrity due to his penchant for recreating famous or iconic images with everyday or found objects (which he subsequently photographs). He is perhaps best known for his use of this technique in Rio de Janeiro's former landfill site, Jardim Gramacho (then the largest in the world), where in 2007–2008 he organized groups of *catadores* or garbage sifters to make copies of the Grand Masters out of items of rubbish found in the landfill. He subsequently auctioned one of the photographs on the international art market in London, with profits returned to the Associação dos Catadores do Aterro Metropolitano de Jardim Gramacho (Allen 2013b: 55–6), and sold others via galleries, thus 'recycling' garbage on a global scale through its transformation, at least at the level of the image, into 'art'. His projects have benefited from wide international dissemination in the art world, and the *catadores* project was the subject of a film documenting the process, *Waste Land* (*Lixo Extraordinário*), made by a UK production team, which appeared in 2010. (The Brazilian title means 'Extraordinary Garbage', a play on the wording that appears on some of the municipal refuse trucks.)

Intriguing and controversial, Muniz's work ultimately relies on an underlying trope of distance, despite the artist ostensibly disavowing the privileged aerial master gaze of his helicopter ride over the vast landfill near the beginning of the film. By this I mean that, no matter how large the 'blow up' portraits of individual refuse workers that result, the emphasis is ultimately on a process of transformation which, as with an optical illusion that requires the viewer to stand at a fixed and distant vantage point to resolve it (one frequently adopted by Muniz in the film), the 'mess' of garbage is miraculously converted into art when seen (and photographed) from the 'correct' perspective. It is true that often the filmmakers of *Waste Land*, Lucy Walker (2010), with João Jardim and Karen Harley, seem concerned, as Allen suggests, 'to neutralise the master gaze of the artist' (2013b: 60) by 'play[ing] on contrasts between proximity and distance' (2013b: 58), moving in and out of the groups of *catadores*. Walker claims as much, vis-à-vis Muniz's aesthetic, in the film's official press notes: 'Vik, as an artist, plays between these levels of proximity and distance, between showing the viewer the material and showing them the idea, revealing the relationship between the paintstrokes and the scene depicted by the paint' (Muniz and Walker 2010: 7).

However, the aesthetic conversion at work in the photographs is ultimately one-way: despite superficial appearances, art is not here disrupted by garbage, in the form of Dadaist shock or the ready-made; rather, garbage is resolved, transcended, or purified in art, and for this reason the work is ultimately normalizing. Dissonance becomes order, poverty is transformed, Midas-like, through the artist's touch, and the Schillerian plot of the *Aesthetic Education of Man*, inaugurating what Jacques Rancière terms the 'Aesthetic Regime' of art (2010a: 116–19), is ultimately reaffirmed by way of a blandly positive answer to Muniz's question 'Can [art] change people?' (Walker 2010: 0:08:17).

The legacy of such charitable hopes that art (especially film) might change the lives of the poor in Latin America, is hardly unproblematic, as Allen points out (2013b: 61), and in a number of cases has ended rather badly, even when it was not an express aim. The most frequently cited example is Héctor Babenco's classic fiction film, *Pixote: A Lei do Mais Fraco* (1982), whose protagonist actor, Fernando Ramos da Silva, semi-orphaned and from a poor lottery-ticket-selling family, was killed in a police shootout after he returned to life on the streets, having failed to make his way as a film/television actor in the aftermath of his lead role in *Pixote*. (In Colombia, Leidy Tabares, the protagonist actor of *La vendedora de rosas* [*The Rose Seller*, dir Víctor Gaviria, 1998] about the lives of street girls in Medellín, whose hopes of an acting career were likewise dashed, was subsequently imprisoned for the murder of a taxi driver in a real-life version of a sequence enacted in the film itself. Back in Brazil, *favela* residents who acted in the 2002 blockbuster *Cidade de Deus* [*City of God*] [Meirelles and Lund 2002] had mixed fortunes, with some, such as Alice Braga, going on to star in Hollywood films, and others arrested for petty thievery, interned in a drug rehabilitation centre, and/or missing, presumed dead [see the 2013 documentary *Cidade de Deus: 10 Anos Depois*].) Walker claims in her blog from the landfill that *Pixote* is one of her favourite movies, noting that her soundman's father was the scriptwriter of the original film (her soundman was José Moreau Louzeiro, and the book *Infância dos Mortos* ['Childhood of the Dead', 1977] by the writer José Louzeiro senior was the basis of the script of *Pixote*). Nevertheless, there seems to be little dialogue with, for example, *Pixote*'s active *disruption* of the false televisual 'cleansing' of urban violence, drugs, childhood exploitation, and prostitution, or rather of their conversion into spectacle. *Pixote* is in fact framed by televisual spectacle: it begins with a scene of horrendous Hollywood violence being watched by a group of young street boys detained in a police station, while in one of the closing sequences,

> the young boy [Pixote] vomits while staring transfixed at the television screen he and his now dead friends had bought with the proceeds of their crimes: what remains of their lives is now a flat televisual surface, and his vomit is the accumulation of all that somehow does not fit within the sanitized screens of Brazilian society.
>
> (Kantaris 2003: 188)

Waste Land and its underlying photographic project inadvertently provide something of a caricature of this critique. The film is, likewise, framed by televisual spectacle – a TV chat show – but the effects of this framing could not be more different from that found in *Pixote*. Muniz's story, the film we are watching, is given to us as his response to a question from the chat show host. Coupled with the rise to stardom of one of the *catadores*, Sebastião (Tião) Santos, depicted on the same chat show at the end of the film, the effect of this is to propel the entire documentary towards the genre of reality TV and its production of minor celebrities as the transubstantiation of life into spectacle, perhaps the quintessential contemporary form of the commodity fetish. Muniz's own 'rags to riches story' (Allen 2013b: 56), which he wishes to replicate with the *catadores*, plays explicitly to the twin global markets of art and film, and provides too facile a repackaging and recommodification of garbage, and of those whose lives revolve around it. This is in fact one version of 'the end of art', where aesthetics finally embraces the full commodification of life, and it is indeed, for Rancière, one of the ways in which Schiller's promise – that 'the art of the beautiful' will transform 'the art of living' – can achieve its full postmodern realization (Rancière 2010a: 116).

In fact there is in Brazil a long and sophisticated tradition of filmic and photographic representations of marginal urban spaces such as the *favela* and landfill sites, together with those who inhabit them, which provides a substantial body of (visual) thought on the representational issues at stake, largely ignored by *Waste Land*. Roberto Stam lists some of these in his essay on the representation of garbage in Brazil, starting with the *Udigrudi* (underground) filmmakers of the 1960s who coined the term *estética do lixo* (aesthetics of garbage) and whose manifesto film was *O Bandido da Luz Vermelha* (*Red Light Bandit*, Sganzerla 1968). As Stam notes, 'For the underground filmmakers, the garbage metaphor captured the sense of marginality, of being condemned to survive within scarcity, of being the dumping ground for transnational capitalism, of being obliged to recycle the materials of the dominant culture' (1999: 70). A related photographic project from the 1970s by Regina Weter played on the minute lexical difference between the Portuguese words for luxury and garbage, with its title, *Luxo/Lixo*, insisting on the structural interdependency of the two apparently antithetical spheres.

But it is in the 1980s and 1990s that filmmakers turn their attention to what the growing piles of garbage produced by Brazil's megacities reveal about the mechanisms that produce inequality on both a national and a global scale. Jorge Furtado's famous 13-minute documentary short, *Ilha das Flores* (*Isle of Flowers*, 1989), follows the voyages of a tomato, as an allegory of the circulation of food-as-commodity, from the vines of a Japanese tomato farmer (in Brazil), via a local supermarket, a housewife's kitchen, her rubbish bin (because the tomato has spoiled), and a refuse truck, to a pig pen, and finally, after the pigs have rejected it, to the women and children scavenging for food on the garbage heaps of Porto Alegre. The allegory does not claim to represent the truth accurately – indeed the film insists in its end credits that it has falsified the place (it is not Ilha das Flores, but Ilha dos Marinheiros), has changed the names of the characters, and it is obvious that the scavenging sessions are staged (although not invented). Nevertheless, the film showed the systemic relationship of garbage to the commodity form and became an instant worldwide success.

Subsequently, the renowned documentarist Eduardo Coutinho turned his attention to the Jardim Gramacho landfill in his 1992 film *Boca de Lixo* (*The Scavengers*, but the title means 'Mouth of Garbage', playing again on *Luxo/Lixo*). In this film, with its close-up shots of crowds of pickers looking for food and other usable items in amongst freshly delivered truckloads of garbage, there is a clear focus on the ethics of filming these subjects, for they argue with the filmmakers and are often reticent to appear on camera until trust has been gained and the human backstory of a number of the pickers is presented. There is, however, no process of filmic sanitization, and the film presents a 'dirty' and uncomfortable image along with a devastated landscape of smoking mountains of rubbish that seem to extend filmically out towards the *favela*-studded slopes of the distant *morros* surrounding Rio de Janeiro:

> Here we see the end point of an all-permeating logic of commodification, logical telos of the consumer society and its ethos of planned obsolescence. Garbage becomes the morning after of the romance of the new. [...] In the dump's squalid phantasmagoria, the same commodities that had been fetishized by advertising, dynamized by montage, and haloed through back-lighting, are stripped of their aura of charismatic power. We are confronted with the seamy underside of globalization and its facile discourse of one world under a consumerist groove. The world of transnational capitalism [...] we see, is more than ever a world of constant, daily immiseration.
>
> (Stam 1999: 72–3)

If garbage is, in these films, linked to the end points of globalized consumerism, it is because the afterlife of the commodity bears the traces of, and is inextricably bound up with, profound mutations in the spatio-temporal coordinates of capitalist production, perhaps at its most intense in global megacities such as Rio de Janeiro and São Paulo. Marx was the first to identify the tendency in capitalism to overcome spatial obstacles to the free flow of commodities through an increase in the speed of transportation, information flows, logistical organization, and turnover. By its very nature, capitalism 'drives beyond every spatial barrier', and the 'extraordinary necessity' it has to overcome distance by increasing the velocity of production, transport, and communication, leads to what he famously termed 'the annihilation of space by time' (Marx 1993: 524). Later spatial theorists, and in particular the Marxist urban geographer David Harvey, call this process 'time-space compression' (1989b: 260–307) and identify it as a fundamental force at work within both urbanization (which shortens the distance between production and consumption) and, later, globalization (which drives beyond the spatial barriers represented by the nation-state system and its fragmented national markets). The megalopolis, which Manuel Castells famously defines as being 'globally connected and locally disconnected, physically and socially' (Castells 1996: 404), in fact condenses both the urban and the global dimensions of time-space compression. Garbage, therefore, does not merely represent the annihilation of a physical commodity, the disaggregation of the raw materials from which it is constructed, the dissipation of the labour locked within its auratic shell and which provided its surplus value, or the final destruction of the commodity's exchange value. As the telos of a form that both condenses and impels the socio-economic state changes through which it passes, it also embodies *the annihilation of space itself*. The apocalyptic dimensions of the garbage dump, as a synecdoche of global ecological devastation, are ample testament in all of these films to the spatio-temporal implosions that the death of the commodity both represents and enacts.

Such perspectives, more or less directly enunciated in the genealogy of films related to garbage outlined above, are the direct precursors of the powerful and complex visions and voices that emerge in what must be the culmination of the Brazilian philosophy of garbage: the film *Estamira* (2004) directed by the photographer Marcos Prado, and enunciated through the hallucinatory and haunting voice of a woman who lived and worked for 24 years in the Gramacho landfill, Estamira Gomes de Sousa.

Estamira: the philosopher of the dump

> If commodities could speak, they would say this…
>
> (Marx 1976: 176)

From 1993, the photographer and filmmaker Marcos Prado began to frequent the Jardim Gramacho landfill to photograph (in black and white) the site and the *catadores*, a project not published in book form until 2004, but for which he received a national photography prize in 1996 (Allen 2013b: 39). Of the choice of black-and-white photography in these still photographs, Allen argues the following: 'In addition to the expansive gesture harboured in the allusion to earlier traditions and values, using black and white also conversely performs a limiting and abstracting function' (2013b: 42). As she goes on to explain, this abstraction, unlike the occasional recourse to aerial photography and high-angled shots in the more famous depictions of charcoal labourers by Sebastião Salgado, does not imply dehumanization, for

> Prado's rubbish pickers, even when gathered together, are always photographed so as to remain distinctly human. Using a telephoto lens in some instances the effect is to bring more distant figures closer, to compress in-between spaces rather than further miniaturise its occupants. This in itself may be said to constitute an ethical decision.
>
> (Allen 2013b: 42)

Instead, here, the use of black and white points self-consciously to a fundamental absence at the heart of photographic representation of the dump and its inhabitants (colour, and perhaps more obviously, smell), and is thus 'a constant reminder of the distance between the image we contemplate and any notion of its capacity to fully capture a certain "authentic" reality' (Allen 2013b: 42).

In 2000, seven years after he first began photography at the landfill, Prado happened to approach one particular refuse picker, a woman in her sixties called Estamira, to ask for her permission to photograph her (Santos and Fux 2011: 128). She not only gave him permission, but told him that she had a vision she wanted to impart to other people: 'She told me she had a mission in life: to reveal and demand the truth [*revelar e cobrar a verdade*]' (Prado cit. Santos and Fux 2011; all translations my own). Thus began a four-year collaborative project with Estamira to film her vision, her philosophy born of the unique perspective of life from the dump, and her life story as told by herself and her closest relatives. The feature-length documentary *Estamira* appeared in 2004, gaining worldwide recognition through festivals and prizes, and a cinema audience of 22,000 people in São Paulo, Rio de Janeiro, and Brasilia after its commercial release in Brazil in 2006 (Almeida 2006). The film is shot in a mixture of digital (DV) and Super-8, interspersing rich colour shots with heavily grained black-and-white, with the frequent switching between these formats preventing the naturalization of either mode and propelling the spectator at times into a spectral world of forms and shapes that are, to use Estamira's own vocabulary, 'disincarnate', and at other times into extreme close-ups of her body, actions, and environment that force us to come to terms with an obstinate material and corporeal world.

Estamira Gomes de Sousa was born in 1941 and died of septicaemia in 2011 (a year before the closure of the Gramacho landfill) in the corridor of a public hospital, unable to get the medical treatment she needed. Although she first began to work at the landfill out of economic necessity, after her husband abandoned her and her children, later, despite the economic support of her then grown-up children, she continued to spend long periods at the dump, partly because of the community she had built up with the other sifters there, partly because the activity, and work in general, gave meaning, as she saw it, to her life, and partly because the dump seemed to give her some profound insights into human nature and society which she saw as her mission to communicate to other people. Estamira uses the film to deliver her theories on society, sexuality, human folly, education, metaphysics, ecology, and religion, with a ceaseless torrent of discourse that is both profound and paranoid in equal measure. Like many so-called paranoiacs, and famously like Freud's Daniel Paul Schreber, she develops highly detailed, fully 'reasoned' and coherent explanations of the workings of society, the universe, and her connection to them. Yet the film does not attempt to judge Estamira, nor does it purport to change her life, nor to 'psychoanalyse' her, although it does investigate her conflictive relationship with the meagre state-provided psychiatric service. It should be noted that at least two psychopathology papers have been written about her case, both of which were based on Estamira's discourse in the film (not on direct contact with her), and both written by members of the Brazilian Congress of Fundamental Psychopathology. The first, 'From Exclusion to the Construction of the World' (Carvalho de Ávila Jacintho 2008) reads Estamira's poetic discourse as a compensation and coping mechanism for the poverty of her material environment. The second, 'Staging Psychosis' (Paes Henriques 2008), a psychoanalytic paper in the Lacanian tradition, more subtly compares Estamira's discourse to Freud's study of Magistrate Schreber (whom Freud likewise never met), to Lacan's *Aimée* case, and to the work of James Joyce, although still with the aim of establishing the parameters for a psychoanalytic (non-psychiatric) treatment.

The purpose of the film – Estamira's purpose – is precisely the opposite of such attempts to frame her as an object of external knowledge, as she tells us in the first words spoken in the film:

> My mission, as well as being Estamira, is to reveal ... the truth, only the truth. Whether it's through the lie, by capturing the lie and rubbing it in your

face [*tacar na cara*], or else by teaching how to see what they, the innocent ones, don't know. Although there are no longer innocents. There are none. There are only 'inexperts' [*espertos ao contrário*].

(Prado 2004: 0:06:40)

Far from the film changing Estamira, as the charitable ethos of *Waste Land* would have it, Estamira wants instead to change us, her spectators, the not-so-innocent inexperts, by rubbing the lies in our face. As Allen puts it, 'it is Estamira's uncompromising judgement that is passed on to the rest of society through the camera rather than the reverse' (2013a: 88). This proposition immediately turns the tables on the entire problematic of 'subaltern representation', as surely as Marx's animated table stands on its head and evolves out of its wooden brain grotesque ideas (see above). This does not of course exempt the film (or this chapter) from the power dynamics of representation, or from the now over-rehearsed debates on testimonial literature and the intellectual construction of the subaltern witness through mediation. But Estamira's violent image of rubbing the lies in our faces is one of contagion by the 'dirt' of the rubbish dump, rather than the conventional politico-theatrical concept of representation within whose logic (and only if we accept the unilateral terms of that logic) the subaltern 'cannot speak'. Even if we often find ourselves questioning the extent to which the film constructs and 'performs' Estamira and her discourse (an inev-itability to which we must of course remain attentive), there is little doubt that Estamira's primary motive in 'making' the film (and indeed her initial challenge to its future director, as cited above) is to invert the relationship of seeing and being seen, of speaking and listening. From the outset, she claims the ground of knowledge and language, and she estranges her spectators' presumed expertise in order to reveal the lies that sustain the existing order of things.

After the above statement, Estamira goes on to give us a synthesis of her philosophy, the unique perspective that living amongst the 'remains' of civiliza-tion has given her. While we listen to her words, the camera roams across the devastated scenery of the vast landfill, picking out details, moving in and out from amongst the garbage pickers, peering at the haze through the flaming methane stacks, tracking Estamira shuffling amongst the huge delivery trucks, or raising our gaze to the vultures and myriad items of flying, storm-tossed rubbish in the skies:

Over there are the hills, the ranges, the mountains … Landscape and Estamira. Esta-sea [*Esta-mar*], Esta-range [*Esta-serra*], Esta-see [*Esta-mira*] is in all places. She is everywhere. Even my feelings see. The whole world sees Estamira. […] This place here is a repository [*depósito*] … of remains. Sometimes they are just remains. And sometimes you also see carelessness [*descuido*]. Remains and carelessness. The one who revealed mankind as the only conditional [*Quem revelou o homem como único condicional*], taught us to conserve things. And conserving things means to protect, wash, clean and re-use as much as possible. […] Saving things is wonderful.

For the person who saves, has. [...] But the Trickster [*O Trocadilo*] made things in such a way, that the less people really have, the more they under-value things, the more they throw away. [...] I, Estamira, am the vision of each and every one of you [*sou a visão de cada um*].

(Prado 2004: 0:8:30–0:15:18)

On the sound track, Estamira's words are accompanied by a rising crescendo from a modern-classical score ('Valse') by Paolo Jobim, son of the renowned Brazilian popular musician Antônio Jobim, from the latter's album *Urubu*, the title of which means 'vulture', or 'turkey buzzard'. While we may balk at the film's apparent aestheticization of Estamira's words, and at the 'sublime' pre-sentation of a post-apocalyptic landscape complete with swirling vultures, a Bra-zilian audience would recognize the ecological significance of Jobim's album from the 1970s (widely considered an early ecological statement) and its reson-ance with Estamira's powerful and purposeful message in this sequence.

Estamira's words and vision overlap in several ways with Benjamin's afore-mentioned allegory of the Angel of History: the pile of wreckage grows skyward; it is a repository of obsolete things and lives, a dire warning of what is left behind in the mad race of capitalist production, and of the apocalypse into which we are rushing headlong; while the collector and recycler is a guardian of broken worlds who would like to make whole what has been smashed, but whose wings are caught in, or perhaps are a manifestation of, the raging storm. They might also agree on an idea of profane revelation ('my mission is to reveal'), although Benjamin's understanding of the 'profane illumination' contains a core of redemptive thought that takes its structure, its *form*, from religious belief. And this is where, I believe, they differ most clearly, for Estamira consistently refuses any messianic meaning or promise of redemption. For her, there is no Messiah, whether figured as revolutionary time, as the *Jetztzeit* (Benjamin 1992: 253), or otherwise. There is only the Trickster (*Trocadilo*), her characterization of (the Christian) god, who plays warped games with people, confuses them, and makes false, misleading promises. Estamira's mission is not a demand for redemption in a veiled theologico-political mode, or on behalf of kingdom-come (figured as communist revolution, Marxist messianism, etc.); it is instead one of revelation in a distinctly profane sense, of lifting the veils that cloud people's minds in the present. She does discuss a concept of communism that she calls '*comunismo superior*', but this notion for her is as simple as it is radical: the practical realiza-tion of full human equality, not understood as homogenization, and not as an equality that is merely an empty or formal right (as declared in Article 5 of the Brazilian Constitution), but one that must be continually *activated* and verified in *practice*, in its *doing*:

All men must be equal, must be communist. Communism is equality [*igual-idade, sic*]. It doesn't mean everyone has to do the same work, or that everyone must eat the same food. Equality is the supreme order [*ordenança*] given by the one who revealed mankind [as] the only conditional. And man is the only conditional, whatever his colour is. I am Estamira, I don't matter. I could be any colour. [...] But I cannot permit, and I dislike anyone offending against colour, or against beauty. This is important: beautiful is what you have done and what you do [*o que fez e o que faz*]. Ugly is what you have done and what you do. [...] Superior communism. The only communism.

(Prado 2004: 1:22:00)

Estamira's political theory is enunciated as she sits amongst the garbage she is recycling, caked in mud, while we are uncomfortably forced to look down from a position of power on a being who simply states human equality, not as one of the Rights of Man, but as the sum of our acts, what you have done and what you do. Humans are the 'only conditional' because our identity, our being, is not given, but is conditional on our deeds, and the Prime Directive of that conditionality is a radical, *unconditional* equality. It should come as no surprise, then, that Estamira's ideas are consonant with those of Jacques Rancière, who insists that equality 'is not [...] a founding ontological principle but a condition that only functions when it is put into action' (Rancière 2004: 52). For Rancière develops his own concept precisely by listening to minor historical voices, such as Estamira's, voices that insisted on challenging, on verifying, formal equality against the myriad limit cases in which it is denied on a daily basis. In a passage that could stand as an elaboration of Estamira's words, he writes:

> To be *intempestive* means at once that you do and do not belong to a time [...]. Being intempestive or a-topian communists means being thinkers and actors of the unconditional equality of anybody and everybody, but this can only happen in a world in which communism has no actuality bar the network framed by our communistic thoughts and actions themselves.
> ((Rancière 2010b: 82) – '*Comunismo superior. O único comunismo*')

Although Estamira sees the phenomenal world of 'odd and even people' (men and women) as a transitory state, and she regularly converses (and argues) with spirits from the 'beyond of beyond' in her own unintelligible tongue, her mission is in no way predicated on a hereafter, whether revolutionary or transcendental: 'I have sometimes wanted to disincarnate! Then I said to myself, "But if I disincarnate now, I will not fulfil my mission." My mission is to reveal, to whomsoever, no matter how much it hurts. And my head toils so much [*trabalha muito*]...' (Prado 2004: 1:42:32). (We shall return to this radical conflation of toil and thought in the final section.) The three phenomena she identifies as most responsible for the veiling or clouding of people's minds are: the fetishism of possessions (possessions that people never really have, even when they think they do); schooling, which teaches people merely to 'copy' and not to see the 'lies'; and, most damningly in her eyes, religious belief, in particular belief in the Christian god. For the first critique, essentially of commodity fetishism and its attendant ecological devastation, she reserves her most haunting visions.

For the second, of education, she feels pity. But for the final, religious critique, she reserves a deep anger and a powerful sense of outrage. Before returning to the first of these critiques by way of conclusion, let us hear her rage and hurt. This is in part driven by her son's belief (as a Jehovah's Witness) that she is possessed by demons and his attempts to have her confined to an institution, and in part by her principled objection to the 'charlatan lies' promulgated by Christianity:

> I have overflowed with rage [*transbordei de raiva*]. I overflowed with rage at being invisible, at so much hypocrisy, at so many lies, at such perversity, at so much Trickster, I, Estamira! [...] They're [doing] the same as Pilate [*Pilatras, sic*] did to Jesus. They have beaten me with sticks to make me accept God. But this god, in this way, this god of theirs, this dirty god, this rapist god [*deus estuprador*], this god who attacks all over the place, everywhere, this god who breaks into people's houses, I will never accept this god. Even if they cut my flesh into a thousand pieces with a knife, with a blade, with anything, I cannot accept it, it's no use. I am the truth, I serve the truth [*Eu sou a verdade, eu sou da verdade*].
>
> (Prado 2004: 1:43:00)

If garbage could speak...

> In the innermost core of the commodity structure there [is] to be found the 'transcendental subject'.
>
> (Sohn-Rethel 1977: xiii)

At the end of the first chapter of *Capital*, Marx imagines what the commodities, which he animated at the beginning of the chapter, might say if they could speak:

> If commodities could speak, they would say this: our use-value may interest men, but it does not belong to us as objects. What does belong to us as objects, however, is our value. Our own intercourse as commodities proves it. We relate to each other merely as exchange-values.
>
> (Marx 1976: 176–7)

If garbage is the commodity stripped of its aura, as suggested above, then we might be tempted to ask '*What would garbage say, if it could speak?*' Estamira, this subaltern, black, impoverished woman, speaks, constantly, unstoppably, throughout the film. She speaks mostly on camera, sometimes in voice-over. She speaks softly and loudly, she shouts, she gesticulates, she swears, she threatens, and she confides. She speaks in lucid Portuguese, she speaks in Cariocan slang, and she speaks in tongues. She speaks on behalf of garbage, for her world view has been shaped by garbage, by this radical perspective that lies beyond the commodity's aura, and by the equality of all things in the spectral afterworld of the dump. This is why the material world is not, for her, 'real'. Instead, she calls it 'abstract', in a passage that conveys, I believe, one of her most profound insights:

> The whole of creation is abstract. All of space is abstract. Water is abstract. Fire is abstract. Everything is abstract. Estamira is also abstract [*Estamira também é abstrata*]. [...] This was always what made me happy: to help people, to help a little creature [*bichinho*]. I have been working here for twenty years. I love it. What I love most is working.
>
> (Prado 2004: 0:16:12–0:18:15)

In order to understand this hallucinatory vision, we must delve a little further into the constitutive role played by the exchange of commodities – the *sine qua non* of garbage – in the very genesis of abstract thought. As we have seen many times, Estamira insists that her mission is to reveal: to reveal a truth that is hidden and that the garbage has enabled her to see. If garbage is a defetishized commodity, what is it that is revealed once the commodity is stripped of its veil or aura?

For Marx, as we know, the commodity effaces the alienated social relations that produce it. It is crystallized labour, but its beguiling appearance suggests the self-sufficiency of a magical object that begins and ends in itself and whose content is veiled. But as Slavoj Žižek suggests in *The Sublime Object of Ideology*, the secret of the commodity is not some content hidden by its form. The secret *is* the form itself, because it is the commodity form that lies at the heart of all social relations under capitalism (Žižek 2008: 3). In fact, the commodity form is an example of what Marx, using a paradoxical phrase, calls a 'real abstraction' [*Realabstraktion*], an abstraction that has material existence as well as material effects. As an object of exchange, it is intimately related to money as the ultimate abstraction that subjects all social relations to the logic of equivalence for the purposes of universal exchange. And for later commentators, in particular for Alfred Sohn-Rethel (discussed by Žižek), who devoted his life's work to exploring this idea, the commodity form that arose through the adoption of coinage can be understood as the real abstraction on which *all* forms of logical abstraction, including most forms of human knowledge, depend.

The analysis of the 'exchange abstraction', of which the commodity form (and money) is the quintessential embodiment, holds the key 'to the historical explanation of the abstract conceptual mode of thinking and of the division of intellectual and manual labour, which came into existence with it' (Sohn-Rethel 1977: 33). It is, in fact, impossible to imagine any separate sphere of intellectual activity without this primary division between intellectual and manual labour, the very basis of class division itself, which the exchange of goods enables, and of which the exchange abstraction is the form. But – and here Sohn-Rethel allows us to understand perfectly Estamira's vision of universal abstraction – the ramifications of the exchange abstraction are even greater, for they extend outwards to transform our entire experience of time and space as abstract entities:

Exchange empties time and space of their material contents and gives them contents of purely human significance connected with the social status of people and things. [...] Time and space rendered abstract under the impact of commodity exchange are marked by homogeneity, continuity and emptiness of all natural and material content, visible or invisible (e.g. air). The exchange abstraction excludes everything that makes up history, human and even natural history. The entire empirical reality of facts, events and description by which one moment and locality of time and space is distinguishable from another is wiped out.

(Sohn-Rethel 1977: 48–9)

If garbage is the commodity stripped of its mystery and removed from the realm of social exchange, then it is, in some sense, the unthinkable and the unrepresentable, the brute rematerialization of the exchange abstraction as inert matter, and the abyss in which the seemingly never-ending chain of equivalences that structure the entire socio-economic realm founders. It is the end point of the logic of 'time-space compression' (Harvey 1989b), as discussed above, that begins with the exchange of commodities and expands through the urbanization, and now the globalization, of capital.

It is, I think, clear now why the chaotic, terrifying, and seemingly limitless commodity graveyard we see in *Estamira* evokes, and paradoxically takes on the qualities of, the sublime, not least for Estamira herself, who finds herself propelled, despite herself, into the spectral realm of the transcendental subject, as the commodities pouring into the dump day after day from the surrounding city rematerialize from the virtual sphere of the exchange abstraction. Surprisingly, this space of absolute exclusion, at the bottom of the social heap, turns out to be one in which intellectual and manual labour, sundered in the act of exchange, become one in Estamira: as her head ceaselessly toils, so she repeatedly declares 'What I love most is working'. For these are the immense forces that are concentrated in an urban waste land that becomes a synecdoche of the ecological devastation of the Earth itself at the hands of the exchange abstraction. In the vortex of the dump, these forces are transmitted, as through a lightning rod, into the fragile frame of an old woman sifting garbage and philosophizing in a forgotten corner of Rio de Janeiro:

The Earth said ... for she used to speak, she did ... but now she is dead. She said that she refused to be a witness to anything. And look what happened to her. [...] The Earth is helpless [*indefesa*]. My flesh, my blood, are helpless, like the Earth. [...] If they burn the whole of space, and I am in the midst, let it burn. [...] If they burn my feeling [*meu sentimento*], my flesh, my blood, if it was for the greater good, if it was for the truth, [...] for the lucidity of all beings, then they can do it right now, this very second [*pra mim pode ser agora, nesse segundo*].

(Prado 2004: 01:45:00)

Note

1 I would like to thank my former PhD student, Alice Allen, for introducing me to the wide variety of Brazilian films dealing with landfill sites and their inhabitants, and in particular *Estamira*. Apart from her work cited in the bibliography, on which I draw, my ideas have benefited from many hours of discussion of this material with her.

5 The paradox of waste

Rio de Janeiro's Praça XV Flea Market

Kirsten Seale

At the end of each day, very little rubbish remains on the streets of Rio de Janeiro's affluent and middle-class suburbs. Through the night and early morning, phalanxes of sanitation workers and scavengers, working in both the informal and formal economies, sort and clean up much of it. Some of that rubbish is handpicked and reclassified as waste, and bound for secondary markets where it can be sold and bought anew (Coletto 2010). Informal and formal second-hand or 'flea' markets are a node within a globally ubiquitous network of secondary economies that generates valuable social, economic, and material infrastructure in cities (Evers and Seale 2014; UNHabitat 2010).

From 1979 until the end of 2013, the Feira de Antiguidades da Praça XV set up every Saturday in Rio de Janeiro, in an otherwise unused channel of land hemmed in on both sides and from above by roadways. The flea market took its name from a nearby square, Praça XV de Novembro, that is both a national monument and a tourist destination. The square and the area occupied by its namesake market are both incorporated in Rio de Janeiro's 'Cultural Corridor', a central urban precinct geographically demarcated because of its heritage and cultural attributes (del Rio and de Alcantara 2009). Following Mary Douglas' (2002 [1966]) influential formulation, the flea market is 'matter-out-of-place' because it is at odds with the official place-image (Shields 1991: 61–2) of historic, touristic Praça XV, and of Rio de Janeiro itself as an egalitarian, modern metropolis (Seale 2014). The market's conspicuous display of waste in the street resists hegemonic projections of what constitutes liveability in urban contexts (Coletto 2010: 59). This, combined with the visible congregation at the city's political, financial, and cultural centre of the market's community of 'urban outcasts' who are usually pushed to the social and spatial peripheries of the city (Wacquant 2008), is interpreted by some as a failure of urban governance (Hiebert *et al.* 2014).

However, counter to the second-hand market's discursive positioning within the representational and material orders of the city, Feira da Praça XV instigates order in an arena where many assume there is none to be found. The market as a space, a set of practices, and a community reinstitutes order amongst previously discarded objects through inventory, exhibition, and above all, commodification. The vendors at the market are entrepreneurial (Seale 2014), reincorporating

Figure 5.1 Feira de Antiguidades da Praça X, Rio de Janeiro, May 2013 (photograph by the author).

waste back into circuits of exchange in a process that provides employment and waste management for the city.

We are socially and culturally predisposed to view waste pejoratively (Elias 1978; Laporte 2000). Some of our rationale for marginalizing it may have a sound physiological basis. Nevertheless, waste is an obligatory, insistent, and above all, valorized component of global, neo-liberal capitalism. Waste is neither abject, nor excessive; rather it sustains capitalism's growth. We might even say, as David Trotter does, that in capitalism 'the success of the enterprise can be measured by the waste-matter it produces, by the efficiency with which it separates out and excludes whatever it does not require for its own immediate purposes' (2000: 22). As indications of the status quo, we can look to the existence of a globalized industry whose driver is the management and movement of the catastrophic amounts of material waste we produce, or to the deliberate configuration of products to deteriorate or to become technologically or stylistically obsolete. To be measured successful, such industries and innovations are dependent on generating increasing amounts of waste. The disconnect between waste's symbolic role and waste's actualized role in global capitalism is what I understand to be the paradox of waste. Through diagrammatic reference to Feira da Praça XV, I aim to construct a theory of waste that acknowledges this paradox.

Waste and refuse

Waste is a convenient catch-all category under which various types of refuse are grouped; for example dirt, excrement, pollution, and garbage. Recent transdisciplinary literature on refuse and/or waste (Hawkins 2006; Scanlan 2005; Hawkins and Muecke 2003) uses the terms more or less synonymously. In theory and in practice the two are often situated as interchangeable, yet they are different. Waste, through its production and consumption, is reincorporated into an economy. Waste therefore has value, whereas refuse does not, because it is no longer circulating within an economy. One of the most cogent explanations of the distinction is by literary critic David Trotter, in his analysis of nineteenth-century Europe capitalism. Trotter says:

> [W]aste can often be recycled, or put to alternative uses; if the system which produced it cannot accommodate it, some other system will. Waste remains forever potentially in circulation because circulation is its defining quality. [...] However foul it may have become, it still gleams with efficiency.
>
> (2000: 20–2)

Trotter's observation on classical capitalism is an iteration of Georges Bataille's metaphorization of biological reproduction as capitalism in *The Accursed Share Vol. 1*:

> The living organism [...] ordinarily receives more energy than is necessary for maintaining life; the excess energy can be used for the growth of a

system; if the system can no longer grow, or if the excess cannot be completely absorbed in its growth, it must necessarily be lost without profit; it must be spent, willingly or not, gloriously or catastrophically.

(1991: 21)

In Bataille's schema, we might understand the excess energy which is reabsorbed to be waste, and the excess which is lost without profit to be refuse.

Waste is legitimized through its exchange value, which (re)constitutes it as a commodity. Waste is therefore neither mere coda to the existence of the commodity, nor its other. Conceptually, empirically, waste and commodity are two sides of the same coin – as the trade at Feira da Praça XV illustrates.

On the other hand, refuse's failure to endlessly reproduce marks the endpoint of commodification, and refuse therefore exceeds the threshold of capitalist requirements. This excess is heterotopic. In the *Order of Things*, Foucault says that heterotopias

[break] up all the ordered surfaces and all the planes with which we are accustomed to tame the wild profusion of existing things, [...] continuing long afterwards to disturb and threaten with collapse our age-old distinction between the Same and the Other

(1973: xviii)

A photographic image of a massive landfill in Mexico by contemporary German artist Andreas Gursky visually represents the heterotopia of refuse. At first glance Gursky's *Untitled XIII (Mexico) 2002* looks monotonous, yet on closer inspection it reveals a chaos of discrete discarded objects: plastics, metals, food matter, stretching beyond the frame of the photograph. There is no order to this profusion, in spite of any attempt to contain it within the architecture of the dump. It represents a promiscuous breakdown of the taxonomies that once ordered all this matter. The commodity's attendant matrix of socio-cultural significations proves arbitrary and unsustainable and becomes redundant when it is discarded as refuse, and the unfathomable scale of refuse accumulating in the dump provides an unwelcome allegory on the fetishism of commodities. We might even think of refuse as the immanent critique of the commodity (Adorno 1973: 97) given, as Karl Marx claims in *Grundrisse*, that consumption, the act which brings the commodity into being, is dependent on the using up and ultimate degradation of the commodity into refuse (1993: 91).

In his 1999 novel *Underworld*, Don DeLillo identifies waste as one of the defining narratives of American (and global) culture going into the twenty-first century. DeLillo writes,

it all ends up in the dump. We make stupendous amounts of garbage, then we react to it, not only technologically but in our hearts and minds. We let it shape us. We let it control our thinking. Garbage comes first, then we build a system to deal with it

(1999: 288)

DeLillo's protagonist Nick Shay works in waste management: 'We designed and managed landfills. We were waste brokers. We arranged shipments of hazardous waste across the oceans of the world' (1999: 89). Like the vendors and ragpickers at Feira de Praça XV, Shay organizes refuse into waste. Waste, unlike refuse, can be 'managed' through reincorporation into capitalist circuits of exchange. The distinction between waste and refuse is therefore not material; instead their classification as one or the other hinges on how we treat them or deal with them. Feira da Praça XV mobilizes this phenomenological difference. The waste that is sold there is other people's leftovers, rejectamenta, detritus. The movement of other people's garbage to the market is also the movement from refuse to waste.

Waste and symbolic and material order

Our notion of what is waste varies from society to society, is dependent upon our material circumstances, and differs according to our position within a social order. Attitudes towards waste can mark cultural variance or similitude. Proximity to or distance from waste is often entangled with the politics of identity. 'Waste', Richard Sennett reminds us, '[is] a problem only dreamed of by scarcity societies' (2008: 109), and assumes a surfeit of resources from which to have leftovers. What ideas and practices to do with waste often share is an objective to sanction, segregate, marginalize, and discriminate. John Scanlan

Figure 5.2 The paradox of waste (photograph by the author).

observes, 'the meaning of waste carries force because of the way in which it [...] operates within a more or less moral economy' (2005: 22). Indeed, waste management as a discourse where the unwanted is first separated and contained and then of use once more is so symbolically powerful that it is metaphorically deployed in a number of social and political contexts (Bauman 2004); for instance work-for-the-dole programmes whose ideological function is to recoup labour capital from 'wasteful' welfare expenditure. In Italo Calvino's fabular city Leonia, waste management performs the sacred task of keeping the unclean at bay. The city's sanitation workers are 'angels' and their task 'a ritual that inspires devotion' (Calvino 1997: 114). Leonia's culture of sanctified waste management absolves its citizens of the consequences of overconsumption. Calvino understands that contrary to waste's symbolic abjection (Kristeva 1982: 65–7), waste is actually a mode of reassurance, because its very categorization is indexical to some kind of order or system at work that sorts, classifies, and divests.

Indeed, in *Underworld*, DeLillo proposes that culture is born of the urge to order refuse:

Detwiler said that cities rose on garbage, inch by inch, gaining elevation through the decades as buried debris increased. Garbage always got layered over or pushed to the edges, in a room or in a landscape. But it had its own momentum. It pushed back. It pushed into every space available, dictating construction patterns and altering systems of ritual. And it produced rats and paranoia. People were compelled to develop an organized response. This meant they had to come up with a resourceful means of disposal and build a social structure to carry it out – workers, managers, haulers, scavengers. [...]

'See we have everything backwards,' he said.

Civilization did not rise and flourish as men hammered out hunting scenes on bronze gates and whispered philosophy under the stars, with garbage as a noisome offshoot, swept away and forgotten. No, garbage came first, inciting people to build a civilization in response, in self-defence. We had to find ways to discard our waste, to use what we couldn't discard, to reprocess what we couldn't use. Garbage pushed back. It mounted and spread. And it forced us to develop the logic and rigor that would lead to systematic investigations of reality, to science, art, music, mathematics.

(1999: 287–8)

DeLillo's assertion is not unconnected to Douglas' assertion that our feelings about the abjection of refuse come out of feelings to do with organization. Douglas observes that we are never more disturbed or repulsed by something than when it has slipped past the processes designed to contain it (2002 [1966]: 5). It is the ruptures or blockages in a system of waste management that produce 'matter-out-of-place'. Waste reverts to refuse. In the city where I live that might be raw sewage washing up on the coastline; or a hoard of rubbish building up at a residence in an affluent suburb (Seale 2006); or a broken, leaking garbage bag

Figure 5.3 Matter out of place (photograph by the author).

lying uncollected on a suburban street. The last example signifies a series of failures. The garbage bag has malfunctioned. The proper (etymologically derived
from *propre*, the French word for clean) place for the garbage is in the bag, not
spilled out on the street. The rubbish and the bag should have been placed in an
appropriate receptacle, or picked up by the city's sanitation workers.

The unease that occurs when waste confronts us with its materiality emerges from expectations regarding order. In *Purity and Danger*, Douglas observes that the desire to install order leads to

> ideas about separating, purifying, demarcating [...that] have, as their main function, to impose system on an inherently untidy experience. It is only by exaggerating the difference between within and without, above and below, male and female, with and against, that a semblance of order is created
>
> (2002 [1966]: 5)

William Ian Miller says that material or symbolic threat to this order provokes the affective response of disgust. Disgust is 'a strong sense of aversion to something perceived as dangerous because of its danger to contaminate, infect, or pollute' (1997: 2; see also Miller 2004). The body, a network of physiological, biological, and neurological processes that constitute a system of order in its own right, cannot cope with a breakdown in the management system, and reacts accordingly. David Trotter elaborates:

> [P]sychological activity [is] an attempt to impose order on experience: bodily paroxysm is a way of confronting and resolving urgent abstract dilemmas. According to this view, you vomit because you have lost confidence in your ability to make sense of the world: your ability to categorize, order, explain, or tell stories about what has happened to you. Disgust is the product of conceptual trauma.
>
> (Trotter 2000: 158–9)

When systems of waste management are functioning as they were designed, the conceptual trauma is allayed, not only through the reversion to order, but because waste management affectively and materially affirms, enables, and finally maintains those aversions.

Feira da Praça XV, as it moved through formal and informal articulations, was always subject to numerous practices of ordering, carried out at multiple micro and macro levels of the market. The infrastructure at the end closest to Praça XV was that of a licensed market. Vendors with more expensive stock in better condition had uniform stalls provided by central management. The regulated parts of the market at this end were more spread out, and consequently were more pleasant spatially and aesthetically. Sometimes, their collections were curated and carefully arranged. This was the closest the market came to the phenomenon of fetishized vintage (Gregson and Crewe 2003; Palmer and Clark 2005). Some were highly specialized in their collections. One stall sold only vintage surplus pencils in their original tins and boxes; another stall was heaped high with (to my mind) sinister-looking used steel surgical instruments. As you moved away from this section of the market, the space under the Perimetral narrowed, and so, too, the space between aisles, stalls, and people was reduced. Here there was a second tier of stalls with red-and-white striped canvas shades instead of green. The wares on display in this section were more eclectic, thrown together, visually messy. Things might be broken, missing a part, or the single remnant of a pair. The flyover's pylons functioned as de facto markers signalling increasingly ad hoc spatial organization and goods, until the market trickled out with groups of sellers who displayed their goods on a towel or directly on the ground. On the fringes of these final sections were groups of Rio de Janeiro's

Figure 5.4 Inventory, exhibition, commodification (photograph by the author).

globally renowned *catadores* (see Lucy Walker's 2010 documentary *Waste Land*) who sorted through the leftovers. Even here, where the market seemed more like a scrapyard than the stage for consumption, there were processes of sorting and ordering at work.

Cleaning up Rio de Janeiro

Like many second-hand markets across the globe, Praça XV set itself up in an otherwise unused urban space; in this case in a wasteland under an elevated roadway in central Rio de Janeiro. Indeed, the market's location and its history are tied up in the making and unmaking of urban wastelands. In the early 1960s construction started on the Perimetral as a measure for Rio de Janeiro's coming automobility. The road cut straight through Rio de Janeiro's impressive Municipal Market. The Municipal Market, which was built in 1907 in the European iron and glass pavilion style with four octagonal towers imported from Antwerp, was subsequently pulled down in stages. The demolition of the central market produced a wasteland in the precinct, by replacing pedestrian and commercial infrastructure with a roadway that cut off the port and harbour from the central business district. Over the next decades, the area's previous function as Rio de Janeiro's central market was metonymically remembered through one remaining corner tower (housing the Albamar restaurant, which continues to trade today), a square named for it on the site (Praça Mercado Municipal), and the emergence of two new markets in the area in the late 1970s: Feira da Troca, effectively a swap meet that functioned through barter, and Feira do Albamar, a more conventional showcase for antiques dealers. The establishment of the markets did not improve the amenity of the area, and the latter moved to the well-heeled suburb of Gávea in the mid-1980s, leaving Feira da Troca to evolve into Feira da Praça XV (Freitas 2013).

The market's terrain within the designated Cultural Corridor, and its inclusion within the catchment area of the *Porto Maravilha* public–private partnership for urban renewal in the Port area, ensured that this space could not remain overlooked indefinitely. The prevailing demands of urban infrastructure development driven by global mega-events – Rio de Janeiro hosted the 2014 FIFA World Cup, and is staging the Summer Olympic Games in 2016 – led to the demolition of the freeway to facilitate rehabilitation of the area. Rumours about the market's relocation dated from at least my first visit in 2008, and were still circulating when I returned in 2013, yet despite the apparent inevitability of relocation some stallholders only found out that the site was no longer available when they turned up one day to find that the dismantling of the Perimetral had begun (Freitas 2013).

Flea markets are often pushed to the social and geographic peripheries of cities because of their trade in waste, their entanglement with informality, and because the communities who patronize flea markets, as producers or consumers, belong to groups who are socially and economically marginalized (Mörtenböck and Mooshammer 2008). The liminal and precarious position within the global urban order of those who work with waste, and of the locations where they carry out their trade is predicated upon social perceptions produced by symbolic representations that locate waste as marginal and abject, not waste's actualized value within local and globalized economies. Informal and/or secondary circuits of commodity exchange, particularly those that manage waste, are indispensable for urban livelihoods in Brazil and globally (Evers and Seale 2014; Coletto 2010), and for socially and materially sustainable cities (UNHabitat 2010), but their vulnerability to the currents of local and global governance and development is compounded by the structural inequalities of a hierarchical informal/formal binarism which deprives those in the informal sector of economic and political power.

The market has since been relocated into Praça XV proper and it remains to be seen if all stakeholders in the market, and in the city more widely, will benefit equally from the 'cleaning up' of the market and its surrounds. The market has been renamed Praça do Mercado Feira de Antiguidades, signalling a separation from the previous market. It is spatially very different, too, in that it spreads out on the square instead of being confined to a linear formation. The increased exposure of this iteration of Feira da Praça XV has not, however, transformed the city's relationship with waste, rather it seems to occlude its presence even more by presenting second-hand goods as a statement of cultural capital or taste, rather than a material necessity or consequence of consumption.

To address the environmental consequences of the paradox of waste requires an epistemic shift in attitudes towards waste, and in the design of associated technologies. Previous epistemic shifts in our social and material relation to waste were designed to render waste invisible. Elias (1978) and Laporte (2000) link the construction of the bourgeois individual to an accompanying privatization of waste. Elias writes that the

> weeding out of the natural functions from public life, and the corresponding regulation or moulding of drives, was only possible because, together with growing sensitivity, a technical apparatus was developed which solved fairly satisfactorily the problem of eliminating these functions from social life and displacing them behind the scenes.
>
> (1978: 137–40)

In her social and cultural history of rubbish in the United States, Susan Strasser retraces how the emergence of new ideas about the correct place for waste instigated the development of industries, products, and consumers for its banishment.

> Personal cleanliness had signified moral superiority among middle-class people at least since the Civil War, and dirt was a sign of degradation. Industrialization made both cleaning and keeping clean easier and cheaper. Cleanliness became big business, as manufacturers of washstands, basins and tubs, towel, plumbing parts, and the large-scale devices necessary for urban sanitation all flourished.
>
> (Strasser 1999: 174)

One reason that the presence of refuse is confronting is that it is a material manifestation of what waste management systems are designed to move out of sight. The visible proliferation of refuse acts as a tocsin warning of unsustainable production and consumption, and resulting environmental degradation. The paradox of waste therefore privileges the symbolism of waste over the materiality of waste. Yet for neo-liberal capital, waste is not a dirty business. Future cities need to transform their relationship to waste in what is effectively an inversion of the current paradox. It is time that city dwellers became far more uneasy with the systems that obscure the material scale of waste production, and far more comfortable with its matter, so that we can interact with it in a way that is realistic and sustainable. Not engaging with waste because it is conveniently hidden away, or because it is considered abject, directly inhibits the design and development of socially and environmentally sustainable and resilient cities.

6   Waste streams and garbage publics in Los Angeles and Detroit

Anne Berg

The 'garbage crisis' that struck public consciousness in the late 1960s continues unabated. The slums of the developing world, garbage dumps like Jardim Gramacho (Walker 2010), children recycling e-waste in China and India, shipwrecks on Bangladesh's beaches, the Great Pacific Garbage Patch – all serve as shocking evidence for impending doom. Critical of invocations of 'crisis', scholars have illustrated that garbage crises have much longer histories and have likely had a consistent place in human imagination ever since people lived in larger settlements (Melosi 2005; Rathje and Murphy 2001). Nonetheless, contemporary rhetoric casts garbage as an urgent global problem, obscuring the deeper histories, placed and raced particularities, and very concrete implications for local people (Frickel 2012; Pellow and Brulle 2005; Pellow 2004; Bond 2002; Bullard 2000; Hurley 1995). Instead, these tales render the places conquered by waste – predominantly located in the Global South – mere symptoms of larger trends and global patterns. But garbage is a global problem only for the 'environmentally enlightened', who do not suffer from economic stagnation, toxic pollution, climate change, soil erosion, population explosions, deadly epidemics, and mass poverty on a daily basis (Mills 2001). Although undeniably bound to and by global capitalism (O'Brien 2008), the global garbage crisis has morphed into a media event of epic proportions against which the Global North performs its techno-political dominance. The ostensible superiority is itself the result of previous garbage struggles that effectively pushed garbage into the margins of Western consciousness. The successful hiding of systemic waste in the West (Gille 2007) enables the affluent, overproducing nations of the world to project their regressive fantasies and real suspicions about absolute limits to supposedly limitless economic growth onto the developing world. Western doom-infused eco-consciousness depends on the piles of seemingly unnoticed, uncollected refuse in 'Elsewhere' to effectively apprise the unaware: if we do not change our ways, this will be our future, too. Yet, even such premonitions have histories. The West's recognition of its own global entrapment parallels its discovery of the environment as a political sphere and the many crises it faces as grounds for political mobilization (Ponting 2007; Dryzek 2013; Allitt 2014).

This chapter focuses on perceptions of the urban garbage crisis in two American cities rather than its subsequent transposition to the Global South. I illustrate that everyday garbage practices invisibly but consistently reproduce the social, racial, and environmental inequalities that pervade (and in fact order) capitalist societies more generally. Retrieving the garbage histories of Detroit and Los Angeles, I tell a triangulated story of crisis, technology, and citizenship.[1] Here, placed histories, struggles over resources, racialized politics, grandiose technological visions, actual materials of history, and their visualization come together and pull local communities and their labour into a matrix that, for lack of a better term, we call globalization.

In the context of their respective garbage histories, the polarities between Detroit and Los Angeles appear in a different light. In both cities, the visibility of garbage in the public sphere has led to multiple recognitions of 'crisis' which, in turn, sharpened political debates over garbage 'solutions' and spawned protests by local citizens. And in each case, garbage problems crystalized as complicated reiterations of past garbage solutions. The notion of a waste stream is crucial here. As a techno-political construct it obscures the underlying systemics of garbage regimes. Garbage does not flow. It is pushed, carted, trucked, and shipped wherever it is ultimately buried or burned. The language of crisis, whether articulated with a local or a global point of reference, reads garbage as a terminal story. In this chapter I am taking the opposite approach, looking backwards, not for origins of the problem but for the origins of presumptive solutions.

Garbage, a visceral thing

One steamy May afternoon, I carry the contents of two campus trash cans to a small recording studio at the University of Michigan and leave it there, in the dark. I am working with a crew to turn a lecture on global garbage into an

Figure 6.1 Michigan garbage, 'dirty' takes (production still).

Figure 6.2 Stench barricades (photograph by the author).

instructional video. Over the next few days, we record the 'clean' takes. Then, in order to compare the university's trash to national garbage statistics, we return to our sample. We set up the light. We spread a tarp to shield the studio floor. I open the bag. A musky, foul, and slightly sweet odour saturates the air. The crew raise their stench barricades. Action. Gloves on my hands and bags on feet, I brave my archive and dutifully separate the soggy papers from plastic bottles, packaging, and food waste. Torn between frivolity and guilt, for about a half hour – the time it takes me to sort, reflect on, and summarize the findings in front of the camera – I feel the power of garbage up close and intimate.

My personal mini-garbage crisis was quickly contained in the same large contractor bag in which it arrived. Later I disposed of our sample in the dumpster outside the building; the carefully separated substances again happily reunited. I (re)acted in seemingly universal and entirely predictable ways. I recoiled from the sight and smell of my sample; I protected the integrity of my own body with plastic barriers and minimal breathing; and, ultimately, I restored a 'safe' distance between myself and the stinking stuff. Mary Douglas's foundational work, originally published in 1966, renders my reaction to the noxious matter emblematic of a learned response to culturally constructed ideas of danger and pollution. Joshua Reno's recent challenge to Douglas's anthropocentric concept of dirt as 'matter out of place' (Douglas 2002 [1966]: 44) goes beyond salvaging the 'thing-power' we feel in the presence of bad matter, instead casting my garbage crisis as a remnant of an animal-typical reaction to 'scat' – a particular and certainly circumstantial interpretation of signs of life (Reno 2014: 9, 15). Whether scat, matter out of place, or both, I got rid of it, leaving behind only a faint whiff of garbage as the studio door slammed shut.

In this chapter I focus on garbage practices and the cultural norms that code social responses to the material presence of waste as natural. Garbage enters public consciousness (and discourse) primarily when the garbage order breaks down – when collection is suspended, when landfill space threatens to run out, when odours or pollutants return to the communal spaces previously liberated from wastes. Michael Shanks, David Platt, and William Rathje (2004: 71) observe that, ordinarily, people remain oblivious to garbage and suggest that 'what sticks in their minds about garbage is the garbage that *shocks* them'. Garbage crises, then, must be understood in light of the particular historical junctures that facilitate or necessitate their respective pronouncement. Such a pronouncement, I suggest, is a political act, a first necessary step toward restoring the essential distance to and anonymity of substances, odours, and pollutants otherwise designated for disappearance.

Rathje and Murphy (2001) suggest that garbage practices have been profoundly resistant to fundamental change and oscillate in varying degrees between four main methods of waste management – dumping, burning, recycling, and reducing the substances in question. As the concentration of garbage increased, methods for its containment and erasure attained political significance. But neither the sanitary landfill – America's preferred and most widely implemented garbage technology (Melosi 2002) – nor the utopian alternatives of zero waste and total recycling (Alexander and Reno 2012: 5) is inherently modern or progressive. Rather, now as in the past, these methods were central to defining and contesting the very notions of modernity and progress as Pauline Goul's discussion of Le Corbusier in this volume underscores (see Chapter 9).

In the nineteenth century, the dual effects of industrialization and urbanization brought about not only a change in kind but in the concentration of garbage. Managing the proximity between people and their waste became a public matter. Historians have illustrated how the growth of cities – the resulting urban congestion, industrial production, and eventually mass consumption – necessitated systematic administrative responses to the overwhelming presence of wastes (Melosi 2005: 9). Increasingly, municipalities relied on the expertise of sanitary engineers to regulate sewage and waste disposal in the name of public health, sanitation, and order (Melosi 2001: 228). This fusion of technological expertise and political authority underwrote modern biopolitical regimes in which 'Each individual has its own place; and each place its individual' (Foucault 1977: 143). Accordingly, waste management played an essential role in the production of social *and* sanitary space that inscribed social, racial, and economic differences as expressions of presumed intrinsic and immutable qualities.

The industrial age not only brought urban populations into the biopolitical folds (Foucault 1977) but it pushed garbage into the public sphere, necessitating solutions that made up for the individual's inability to reliably channel unwanted substances elsewhere. City dumps, incineration plants, sanitary landfills, and materials recovery facilities were certainly not the civic spaces Habermas (1989) had in mind when conceiving the congealing bourgeois public. However, it is precisely the publics of civil servants, Nimbyists, sanitation officials, environmental activists, politicians, waste workers, and everyday trashers that have captured my interest – not as the authoritative voice public opinion (*öffentliches Räsonnement*) (Habermas 1989: 28) but as a rostrum on which personal grievance performs politics.

Being modern

A fine mist of enzymes and essential oils thickens the air and imperfectly masks the atrocious stench of three-thousand-and-some tons of raw garbage on the tipping floor at Detroit Renewable Power.[2] Front-end loaders move the material onto arrays of conveyer belts to be visually screened for hazards such as propane tanks, mattresses, and car batteries. The waste is then shredded twice over, mechanically stripped of metal content and stored as refuse-derived fuel (RDF). Combustion of RDF turns water into steam, which powers a turbine generating electricity to service 68,000 homes. The remainder of the steam is used to heat and cool the city of Detroit via an underground system of pipes. Outfitted with a bag house and tightly monitored for emissions, the facility is nonetheless unable to hide its age. The incinerator's potential for and history of protest is not only readable on countless websites lambasting the plant, but also in the comportment of the few onsite managers. The odour-binding mist is unable to hide that, underneath, garbage continues to stink.

Figure 6.3 Detroit Renewable Power: the furnace (photograph by Rory McGuinness).

Currently, the incinerator is entangled in a class action suit over odour viola-tions.[3] Earlier this year Zero Waste Detroit, a coalition of community and environmental groups, insisted on the installation of better odour control techno-logy, explaining that noxious odours prevent nearby citizens from opening their windows (Zero Waste Detroit 2014). Citing adverse effects on property values and diminished use-value of the urban environment, current opponents have shifted their language registers to keep up with the tenor of neo-liberalism. But controversy has surrounded the incinerator ever since it moved beyond the plan-ning phase in the mid-1980s. Then, enraged citizens and environmentalists in Detroit found militant support three miles across the river – in Windsor, Canada. In September 1986, the Province of Ontario joined the Environmental Protection Agency (EPA) in legal action, demanding that Detroit install state-of-the-art air pollution control equipment, which the original designs for the plant forewent. Since environmental standards were more stringent in Michigan than at the federal level, the city proposed to build the incinerator in accordance with federal standards. In 1984, the city obtained a permit to commence construction without said pollution controls, which would have cost $17 million to install and an additional $8 million annually for maintenance purposes. By October of 1986, the EPA abandoned its legal efforts to revoke said permit after the Federal Court ruled that the EPA had missed opportunities to intervene before the permit had been granted. The plant finally opened for business in October 1988. An Inter-national Joint Commission ordered by the US and Canadian governments began studying the relationship between emissions and health effects in the Detroit-Windsor Area and in subsequent months community activism intensified considerably. In April 1990, the incinerator was shut down intermittently for violating mercury emission limits, and a year later the city sold the plant to tobacco giant, Phillip Morris. In 2010, Atlas Holdings and the Ohio-based Thermal Ventures II formed Detroit Renewable Energy and purchased the incin-erator, which is now operated by a subdivision, Detroit Renewable Power. The plant remains within emissions standards. But to ensure that the waste-to-energy (WTE) plant operates profitably, Detroit imports garbage from wealthier sur-rounding areas, many of which are also predominantly white. Only 340,000 tons of the 800,000 tons the plant combusts in a year come from Detroit. In 2014, the city still lacks systematic kerbside recycling and environmental activists in Detroit blame the incinerator.

To locate the incinerator at the source of these controversies would be a sim-plification of much more complicated historical continuities. The WTE facility was itself born out of controversy and crisis. On 1 July 1980, city employees went on strike as the city refused to meet their demands of an increased base-pay and cost-of-living improvements for municipal workers. Parks closed. Bus ser-vices ground to a halt. Garbage piled up and festered in the summer heat as if to greet early arrivals to the Republican National Convention. Sanitation workers had achieved a substantial victory in a four-day strike in 1971, negotiating for a cost-of-living allowance, better pay, health and retirement plans, and, most sig-nificantly, the recognition of the birthday of Dr Martin Luther King, Jr, as a paid holiday. In the summer of 1980, 11 days passed and 55 million pounds of garbage accumulated in the city before workers ratified a new contract. When the Republicans convened the following day, there was little doubt regarding the power of garbage. The national press covered the Detroit garbage crisis as a symbol of union brawl and urban decline.

In this context, the idea of a waste-eating plant presented itself as a modern solution to a presumably modern affliction. But turning trash into revenue via the detour of steam was neither a quick nor a cheap fix. When Mayor Coleman Young announced the project in mid-September 1981, the city estimated that the construction of the incinerator would total about $300 million. By the time the next garbage crisis struck Detroit in 1986, construction was already underway and the projected grand total had risen to a stunning $420 million. The stakes were also considerably higher. Detroit's first black mayor was in his twelfth year. Many hopes for Coleman Young had been disappointed and he (unfairly so) bore the stigma for Detroit's decline (Reed 1999: 81).

On 16 July 1986 municipal workers walked out on the city government. The strike, the longest in Detroit's history, lasted 19 days. About 7,000 city workers were actively on strike. Another 5,000 workers honoured the picket lines. Temperatures were in the nineties. The city became a smorgasbord for rats as 76,000 tons of garbage remained uncollected. Makeshift dumps sprang up across the city. Downtown businesses contracted private haulers to keep their district relatively garbage free. More-fortunate Detroiters paid their poorer neighbours a few cents to lug their household waste elsewhere. Garbage accumulated unchecked and most rapidly in the poorest parts of the city. And yet, though plagued by flies and odours, citizens remained remarkably upbeat. Detroit, some exclaimed, is no stranger to adversity.

As the poor knew all too well, Detroit's crisis was not garbage born. Rather, garbage was one of the creeping crisis' most visible symptoms. The auto industry was steeped in a recession and the city spiralled deeper and deeper into debt. White flight had been a constant trickle ever since the wartime booms led to a labour influx of southern migrants who hoped to forge their own luck in Motown. White hostility increased exponentially after the revolt of 1967 and with Young's election to mayor, the trickle turned into a river over the course of the 1970s, leaving behind a predominantly black city by the early 1980s (Thompson 2004: 206). White flight stripped the city of its tax base and contributed to rapidly growing unemployment rates among the remaining urbanites (Sugrue 2005). Unemployment in turn transformed the urban geography. Derelict houses, blight, and poverty spread across Detroit at the same time the downtown experienced its first round of redevelopment and gentrification (Thomas 2013: 157–60).

When environmentalists today decry the incinerator not only as the epitome of urban decline but as an embodiment of environmental racism, they are only half right. It is true, the incinerator is located in a predominantly black residential area. It is further true, that trash in Detroit is predominantly handled by black workers. Even before the mass exodus of white citizens, 90 per cent of garbage workers in the city were black (Walsh 1975). But incineration was, at least initially, understood as an emblem of modernity supposed to extend the promise of urban domesticity, cleanliness, efficiency, and convenience to wider segments of the population. In the 1980s, such goals already had a long history.

In the first decades of the twentieth century, garbage, particularly when irresponsibly dumped, left behind, or accumulated in streets and alleys, was already seen as a major problem demanding modern technological and technocratic solutions (Lovejoy 1912). In Detroit, urban planners, administrators, and businessmen considered it their duty to maximize the use of resources and extract value from waste. Although Detroit lacked a clear system of garbage disposal and collection before 1920, the city was immaculate according the Baltimore Commissioner of Street Cleaning, who examined Detroit in 1900. Throughout the first half of the century, Detroit was emulated as a model of efficiency and productivity. With respect to waste management the city systematically moved toward more technologically and administratively complex strategies.

As early as 1912, every house, dwelling, and building in the city of Detroit was to provide a sanitary receptacle with a watertight lid in which all offal, including animal and vegetable matter, was to be collected after having been thoroughly drained. Since garbage (mainly organic substances) and rubbish (paper and other combustibles) were often collected separately, different disposal methods developed for each. For mixed refuse, incineration was often seen as the most effective and most modern method of disposal, advocated by Detroit's Board of Commerce as early as 1911.

Incineration was not the only industrial method for extracting value from waste. Reduction, as its main competition was called, required a factory-like plant in which garbage (putrescible refuse) was stewed in large vats to separate oils and glycerine from the 'residuum'. The former found various applications in industry, for example as a machine lubricant, and the residuum was sold as fertilizer (Melosi 2005: 159), the liquid runoff was channelled into rivers and streams.

In industrial centres with a large demand for machine grease, reduction was also particularly appealing. In 1929, Henry Ford offered to take all of Detroit's garbage as well as that of its neighbouring municipalities for processing in a massive reduction plant in Dearborn. The Depression motivated Ford to systematically eliminate waste from the production process and Ford's Industrial University – as the Edison Institute for Technology in Dearborn was called – experimented with the use of garbage to correct for civilization's waste of nature's resources. Ford used up to nine tons of garbage daily to produce a new kind of hyper-potent fertilizer and distilled commercial alcohol as well as various oils from organic refuse. Moreover, Dearborn became the nation's foremost pioneer in sewage sludge incineration, eventually burning every pound of sludge produced by sewage treatment plants in the city (Owen 1938). Evidently, incineration and reduction operated side by side, lauded by engineers and health officials as highly preferable to dumping.

By the middle of the century, 254 truck drivers and 508 garbage pickers effectively serviced Detroit at the cost of $3 million per annum. Hog feeding, landfilling, combined sewage-garbage disposal by way of industrial garbage grinders, and incineration were used throughout the region, but in Detroit, where wet garbage was collected weekly and combustible rubbish biweekly, incineration remained prevalent until after the Second World War. As long as Detroit was a centre of industrial production, municipal services and industrial rationalization entered into an almost symbiotic relationship – even the police played its part by transporting abandoned cars to the Detroit House of Correction where they were gutted for scrap and demolished.

Modern efficiency, however, was not only of interest to industry. Since before the Great War private home incinerators were considered luxuries akin to automobiles. These devices burned combustible refuse inside the home or, a cruder variant, in backyards. Home incinerators and garbage grinders essentially offered households the ability to disappear the wastes in private, mimicking but not depending on industrial processes. Air quality was certainly a problem in the 1950s. The city's Air Pollution Control Program, initiated in 1947, was a rather unique organization that worked closely with representatives from city and industry, monitoring air quality, inspecting facilities and mandating technology updates. By 1953, $14 million had been spent on upgrading pollution controls, not including the voluntary installation of pollution control equipment. The results, according to *Public Health Reports* (Linsky 1953), were quite remarkable, with significant increases in horizontal visibility and reduced dust fall in industrial areas. Fly ash no longer accumulated in the streets with more than 140,000 tons now being captured.

By the late 1940s and early 1950s home incinerators looked like washing machines or freezers. The Detroit Jewel, made by the Detroit-Michigan Stove Company, was a top-loaded machine with controls that allowed for the adjustment of heat depending on the nature of the waste. The ash, so advertisements suggested, could be used for fertilization of home gardens. Gas- and electric-powered home incinerators, 35,000 of which were installed in Detroit in 1953 alone, were most prevalent in white middle-class households, and together with garbage grinders effectively rid citizens of unwanted substances. The massive increase in home burners in the 1950s cannot solely be explained as a function of convenience and overconsumption that characterized the throwaway society of the decade (Strasser 1999). Surely home burners and garbage disposers offered convenience and modern gadgetry that conveyed status and effortless efficiency. They also guaranteed at least a modicum of disposal independence. In light of the garbage strikes that had disrupted life in metropolitan Detroit in 1943 and 1950, the home incinerator liberated middle-class families from the strike patterns of the city's largely black army of garbage collectors, working class solidarity, high levels of unionization across industries, and a strong union presence in municipal politics (Sugrue 2005; Thompson 2004; Lewis-Coleman 2008; Bates 2012).

Detroit's decline, though most notable in the 1980s, in fact had its roots in the racialization during the city's wartime boom. Since the growth of the auto industry in the early twentieth century, Detroit saw a massive influx of southern migrants (both black and white), in particular after the Unites States halted European immigration during the Great War (Bates 2012: 16–17). During the second Great Migration in the 1940s, five and a half million blacks from the rural south hoped to make a fortune in America's 'arsenal of democracy' (Thompson 2004: 7). Detroit was 'one of the nation's fastest growing boomtowns and home to the highest-paid blue collar workers' with a very strong union presence and thus not only promised job security but possibly social mobility (Sugrue 2005: 3, 110). Though unionized, black citizens were far from integrated – neither on the job and much less in residence (Lewis-Coleman 2008). Starting in the 1930s, the auto industry took its first steps to defuse union power and combat wage hikes by decentralizing the workforce and moving production out of the city centres. While these early attempts did little to transform Detroit, they anticipated the industry's post-war efforts. In the 1950s, the disaggregation and reduction of the urban workforce dramatically reshaped the industrial geography (Sugrue 2005: 128). The effects were particularly devastating for the city's black workers. Lower in seniority than their white counterparts, African American workers were usually the first to be fired. Moreover, black workers often held the dirtiest and least pleasant jobs whether in the auto industry or the public sector. Of the African American Detroiters who worked for the city in the 1940s (and roughly 36 per cent of municipal workers were black) large numbers toiled as janitors, sanitation workers, garbage men, and heavy labourers. Many were seasonal workers in the city's parks and public works; such jobs still bore the stigma of New Deal work-creation programs (Sugrue 2005: 110–11). In the auto industry, automation replaced the less-skilled jobs first, and since discrimination had ensured that such jobs were predominantly held by black workers, decentralization and automation combined with discriminatory hiring and firing practices to produce an under- or unemployed urban underclass of predominantly black citizens, which in turn reinforced racial stereotypes of their white counterparts (Sugrue 2005: 144).

Racial discrimination was not limited to the job. As Thomas (2013) illustrates, residential segregation was vigilantly enforced by white residents. But it too had its roots in the 1940s. By 1944 more than half of the city's African American population lived in overcrowded, substandard dwellings in the city centre (Thomas 2013: 17). Housing outside the congested urban core was not open to people of colour and violent reactions to black mobility by white neighbours escalated in the riots of 1943. In the aftermath, subsidized, low-income housing became formally based on segregation until this policy was rescinded in 1952. Such a reversals in policy hardly altered white attitudes, as evidenced by the continuous exodus of white citizens. The slum clearance and urban relocation projects of the 1950s increased segregation, destroyed existing communities and escalated pressures on already overcrowded black residential neighbourhoods (Thomas 2013: 56–64). Over the next decade racial tensions exacerbated. The silent consensus that underwrote redlining and residential segregation was constantly reinforced by police brutality, racial discrimination, and racial violence, a volatile mix that exploded in the summer of 1967 (Thompson 2004: 97–102).

When Mayor Coleman A. Young assumed office in 1974, the situation was dire to say the least. Young aggressively pursued an agenda of urban development, forged alliances with big business and hoped to eliminate blight and unemployment (Thomas 2013). A year later, the city faced the worst fiscal crisis since the Great Depression. Mass layoffs and plant closures left Detroit to struggle with astronomical unemployment rates (Thompson 2004: 207–8). The national economy certainly suffered in the aftermath of the 1973, but the oil crisis devastated the auto industry and thus disproportionately affected Detroit. In this context, it is hardly surprising that WTE plants experienced a comeback. By the early 1980s as landfill space threatened to run out, many cities looked to incineration as the technology of the future. New York proposed to build five WTE plants by the early 1990s. In Detroit, plans for landfills turning into golf courses, ski-resorts or amusement parks now belonged to the past.

In 1989 Detroit had the highest unemployment rate in the United States and the lowest median income in comparison with peer cities (Thomas 2013: 175). In that same year, a black mayor built a municipal incinerator to extract profit from the city's garbage. The logic of technological modernity clashed with the raced concepts of trash. Charles W. Mills has convincingly argued that 'for centuries in the United States, blacks *themselves* have been thought of as disposable' (Mills 2001: 74, original emphasis). And even though many talk about the garbage crisis as 'an environmental challenge for an undifferentiated raceless "human" population', this challenge is disproportionately born and managed by society's 'sub-persons', who are not considered part of the 'we' the white majority invokes when asking about 'what do to do with our refuse' (Mills 2001: 73, 84). Rather, black Americans continue to be constructed as 'an environmental problem' in its own right that is nearly congruent with the problem of waste (Mills 2001: 84). Detroit's black population remained loyal to Young and the garbage plant threatened to spew the pollution of poverty, blight, and blackness into the surrounding communities. When the predominantly white, educated, young, and leftist students, professionals, and residents from the Cass Corridor – together with their Canadian allies and local environmental groups – attacked the incinerator as a symbol of the many failures of global capitalism, they lacked the language of environmental racism that their successors in 2010 and 2014 deliberately mobilized (Thomas Stephens 2014, personal conversation). In 1989, the majority of African Americans in Detroit continued to support Mayor Young and his decision to build and operate the incinerator. Environmentalists spoke for themselves and on their own behalf rather than in the name of social justice. Armed with the language of universal human responsibility, they had entered the post-racial fantasy land, in which the only colour that matters is green.

The garbage public

On 31 October 2013, the day the Puente Hills Landfill closed for good, an official from the Sanitation Districts of Los Angeles County approaches the SUV of Basil Hewitt, who is slowly driving toward the day's open cell. 'A worker spotted a journalist, there's a picture-taking journalist is on the premises', the official informs the senior engineer, breathlessly. However, before he finishes the sentence, his eyes come to rest uncomfortably on the Canon D60 on my lap. 'She is a historian', Hewitt explains calmly to the man staring suspiciously through the open car window. I continue to take pictures. I did not get to observe the very last truck that dumped its cargo at the country's mother of dumps, but I came close enough. Hewitt, who generously guided me around the complex and patiently answered my questions, gazes across to Roseville Memorial Park Cemetery and mumbles something about good neighbours. In contrast, the growing residential community of Hacienda Heights – faced with the state's largest landfill in their backyard – quickly discovered their inner environmentalism.

Mike Davis (1992) has illustrated the significance of property, homeownership, and homeowner's associations in shaping the social geography of the larger LA region. Homeowners, he argues, acted as self-styled *'sans culottes'* in defence of the social and racial exclusivity of their neighbourhoods (Davis 1992: 156). In Los Angeles, the propertied mobilized with equal force against garbage as they did against multifamily housing. These civic institutions not only promoted the fragmentation of the metropolitan area and its racial stratification, they

Figure 6.4 Trucks line up at the entrance to Puente Hills Landfill on the last day of operation (photograph by the author).

were also at the forefront of the new urban environmentalism that pitted home-owners, park agencies, and well-healed environmentalists against county offi-cials (Davis 1992: 170). After Mission Canyon Landfill closed in 1965, the city facilitated the building of private schools, luxury homes, and religious institu-tions to claim the rim of the beautiful canyon in the Santa Monica Mountains. When the Sanitations Districts proposed to reopen Mission Canyon in 1977 and site additional landfills in this exclusive corner, the posh communities of north-western Los Angeles organized in protest. Although the proposed sites were only to receive a fraction of the 38,000 tons of waste the region produced each day, homeowners – worried sick about property values – raised a stink, citing air quality, odour problems, groundwater pollution, and wild life preservations as primary reasons for their opposition to landfilling in their vicinity. Mission Canyon was never reopened and the property values in the Santa Monica Moun-tains continued to soar.

When the garbage crisis struck Los Angeles in the 1980s, it was in part the result of these very movements and their institutions. In 1980, Los Angeles still had 19 landfills, but their numbers were shrinking rapidly, as existing sites reached capacity, permits expired, renewals faced vociferous protest, and citizens effectively blocked the siting of new landfills. Meanwhile, the Califor-nia Waste Management Board (a state agency) and Mayor Tom Bradley, Los Angeles' first and only black mayor, spent $500,000 to convince citizens that the larger LA area was in fact facing a garbage crisis, organizing a rally on the steps of City Hall. This rally drew less than a dozen people. Mike Selna, an engineer with the Sanitation Districts, explained, Los Angeles was not facing a 'trash crisis'. Instead, the city was facing a permit crisis. The resulting domino effect increased pressure on existing landfills, such as Puente Hills, but also forced a revolution in the city's garbage practices by the early 1990s, when Los Angeles institutionalized the nation's largest mandatory recycling programme. Initially, however, the recognition of the city's daily garbage tsunami neither produced a sense of solidarity nor forced political solutions to a shared legacy. Instead, the garbage crisis of the 1980s inspired localized movements that insisted on placing as much distance as possible between the regional waste stream and their properties, a vision that ultimately prevailed. The subsequent commitment to environmentalism and mandatory recycling did not change the desire for distance. Rather, Los Angeles reimagined this distance in explicitly 'green' terms.

The story of Puente Hills is illustrative of Los Angeles' larger garbage dynamics, and it too ends with recycling. The landfill started out as privately owned dump in San Gabriel Valley, which then had plenty of unused, open land and was still conveniently close to significant populations (Humes 2012: 51). By 1980, Puente was operated by the Sanitation Districts and served as the state's busiest landfill. It accepted roughly a quarter of Los Angeles' garbage – up to 13,000 tons daily. Buffered, lined, and outfitted with its own power plant to utilize landfill gas (even including a gas station for employees), the facility was landscaped to seamlessly integrate into the surrounding countryside.

Figure 6.5 Landscaped garbage mountain, Puente Hills (photograph by the author).

Perhaps it is not a coincidence that Puente opened the year that home incineration was banned in Los Angeles. In 1957, more than 1.5 million backyard incinerators were shut down in Los Angeles County. The new law not only prohibited the burning of trash but also provided for municipal trash collection in Los Angeles and 15 other cities. Private haulers, and many of them small, were still the norm and continued to provide the brunt of garbage collection services in the larger LA area as they had (and would continue to do) for much of the century. Concerns over air quality justified the ban on incineration but now necessitated the expansion of regular garbage collection in the county. Nonetheless, large areas, including 13 cities, found themselves without formalized refuse collection. By the end of the 1950s, only about 90,000 households served by the county Sanitation Districts enjoyed regular municipal services. In contrast, 270,000 in county territory were provided with the names of private collectors instead, required to make individual arrangements.

In Los Angeles, waste management developed to accommodate a 'new kind of industrial society where Ford and Darwin, engineering and nature, were combined in a eugenic formula that eliminated the root causes of class conflicted and inefficient production' (Davis 2001: 97). Presiding over the city's explosive growth in the early twentieth century, an elite of wealthy outsiders transformed the region into an industrial powerhouse. By the early 1920, Los Angeles eclipsed San Francisco's industrial capacity, and over the next 20 years, morphed into a regional metropole and a global city (Ethington 2010). With an economy based on citrus, oil, motion pictures, and soon aircraft manufacturing, Los Angeles constituted the centre of the nation's military-industrial complex and successfully transcribed the open-shop ideology of its oligarchy onto the urban grid. Segregated, heavily policed, and staunchly anti-unionist, the metropolitan region valorized private property, industrial freedom, political spectacle, and entertainment and embodied the high-modernist production ethos of Cold War America (Ethington 2010).

The development of waste management and attitudes toward garbage not merely reflected the industrial dynamics of the Los Angeles region but channelled regional ecology. In 1905, a mechanical engineer from Newark studied this nascent metropolis for several months and was baffled by the city's wastefulness. With the right technology, he calculated, the city could extract revenue of roughly $3 million from garbage by producing gas and fertilizer instead of spending between $70,000 and $100,000 annually to get rid of it. But abundant natural energy – sunlight and oil – hardly provided an incentive to emulate East Coast and European attempts to extract energy from waste. In the eyes of Angelenos, incineration threatened to destroy property and reduction plants were uneconomical. Only agriculture provided a venue for utilizing putrescent garbage by feeding it to swine. But after the hog cholera outbreak of 1911, Los Angeles county banned the feeding of hogs on city garbage to protect the health of county citizens. When the Pacific Reduction Company, which had taken the place of hogs farms in the meantime, threatened to close in 1921, the city urgently looked for solutions to deal with the 250–300 tons of garbage Los Angeles churned out daily. The solution to the 1921 garbage crisis anticipated future solutions. By the end of the summer, the city entered into a contract with Fontana Ranch, an industrial hog farm in San Bernardino County, where 20,000 swine devoured the cargo that was now shipped across county territory in enclosed railroad cars.

The struggle between city and county over refuse proved long-lived. The region's growth and the political solutions implemented to manage it only exacerbated this dynamic. The incorporation of Lakewood in 1950 set an important precedent for the proliferation of incorporating municipalities between 1954 and 1970. As the metropolitan region swelled with non-white migrants, Los Angeles experienced a mass exodus of whites. Incorporation offered protection of property, racial exclusivity, home rule, and low taxes to white middle class communities that regrouped in the suburbs, while the burden of providing municipal services was outsourced to the county (Ethington 2010: 204–6; Hogen-Esch 2010: 236).

The ethos of private property also reordered garbage removal. In the 1930s, citizens groups and the League of Women Voters launched a campaign to relocate residential dustbins to the back of each house so as not to soil the facade of their neighbourhoods. By 1950, the city encouraged the installation of home garbage grinders to reduce the need for residential dustbins and collection, eliminate the practice of hog feeding, and improve the overall aesthetics of the metropole. At the end of the year, the garbage again invoked a sense of doom as San Bernardino and Orange Counties adopted ordinance to prohibit the feeding of garbage to swine. Fontana Ranch, potentially the largest garbage-to-pork factory in the country, closed the following year. Incineration was categorically ruled out; cost and odour issues being the most frequently cited reasons. The city remained ambivalent about landfilling and continued to advocate rail transport of garbage to Elsewhere.

When the garbage crisis returned to Los Angeles in the 1980s, landfilling had long since become the dominant method of disposal. Certainly, garbage and sanitation workers went on strike in Los Angeles as they did in other cities, but due to the disaggregated system of collection by many small and medium-size private firms, such strikes did not have the potential to lay waste to the city as they did in East Coast and Midwestern municipalities. Here, the garbage crisis materialized as a result of landfill space running out. The garbage public of enraged homeowners ensured that no new landfills were sited in proximity to their properties. Instead of insisting on higher wages and health benefits, like the marginalized labourers in eastern towns, homeowners learned to mobilize against landfills and incinerators in the name of environmentalism – soon a ubiquitous surrogate for property values.

Over the course of the 1980s, the middle class community of Hacienda Heights around Puente Hills Landfill and even Republicans such as State Senator William Campbell became outspoken 'environmentalists' and mobilized together with homeowners' associations against the public health (and voting booth) hazard – the dump. They packed public hearings with hundreds of people vehemently opposed to extending the fill's lifespan over an additional 29–50 years and expand its capacity by 106 million tons. Citizens remained unmoved when proponents of extension explained that the closure of Puente Hills would amount to 40 per cent net loss of the region's dumping capacity and leave 60 cities without viable alternatives. Sanitation Districts officials estimated that, if Puente were to close in 1983, the county would run out of landfill capacity by 1992. Residents around Puente, such as Susan Waite, lamented the destruction of nature, of oak and sycamore trees, while thinking about the value of their homes. In 1983, the Sanitation Districts won the battle. Pressures by the Los Angeles waste stream were simply too great. But the fight continued alongside the search for alternatives.

Incineration briefly appeared as a potential solution. The Sanitation Districts envisioned a WTE plant at Puente Hills landfill that would power 500,000 homes and burn 10,000 tons of trash a day. At the same time, plans for an adjacent recycling centre were under way. But the citizens of Hacienda Heights were to have neither. Two years later the project was dead. Additional WTE facilities were proposed in Azusa, Irwindale, South El Monte, and Pomona and here too, the projects all failed because of community opposition. Yet the fight over the dump's expansion dragged on. In October 1990, opponents filled a school gymnasium insisting that the people of San Gabriel Valley had done their share. The stuff should be shipped to Elsewhere, far, far away.

In the fall of 2013, Puente Hills landfill finally closed, years before reaching capacity. The 'solution', proposed numerous times through Los Angeles' garbage history, would become a costly reality. Shipping waste 220 miles inland became the only politically viable resolution in light of community vigilance and the growing amounts of municipal solid waste. Years prior, the Sanitation Districts had purchased and developed Mesquite Regional Landfill in Imperial County as the designated site for the implementation of a massive waste-by-rail project to guarantee 100 years of disposal capacity. Initially the Sanitation Districts had targeted two sites – Eagle Mountain and Mesquite. Eagle Mountain citizens won a legal battle to ward off the 20,000 tons of Los Angeles garbage per day. Mesquite Regional Landfill is getting ready to start operation as soon as existing landfill space in the greater LA area is exhausted. Located in the desert, the $36 million project is far from groundwater, people, and homeowners' associations (at least for now).

For the past decades, Puente had been the designation for roughly a quarter of Los Angeles' waste stream. In 2005, a materials recovery facility (MRF) opened adjacent to Puente Hills Landfill sorting 4,400 tons of trash per day. The Puente Hills power plant will burn landfill gas for another 30 years and the trucks will keep coming. The MRF continues to operate and will be a central node in the waste-to-rail project. In the meantime, the stuff that cannot be recycled is loaded onto trucks and transported to privately operated landfills across the county where residents are less vigilant, where single-family homeownership is less prevalent, where communities are less affluent and less white. Since the landfill's closure, the Sanitation Districts added a 'dirty' MRF where a mainly female minority workforce – subcontracted through a private agency that fails to provide benefits – manually removes recyclables from regular household waste before the latter is loaded onto trucks or eventually rail wagons. America's largest and one of its greenest cities pushes garbage further and further down the social ladder and out of sight of affluent citizens. Whether the prices of homes will rise to counterbalance the increase of garbage fees for citizens of Hacienda Heights remains to be seen. The garbage *Öffentlichkeit* has spoken. The sycamore tree has been saved.

Elsewhere

The parallel garbage histories of Detroit and Los Angeles toy with our categories of modern, industrial, progressive, and green. The determination to recycle undergirds not only our political fantasies about 'zero waste' and global sustainability but also redeems our desires for modern comforts. Such technologies now proffer the salvation of the ailing planet, forging a global consensus among those on easy street. For the first half of the twentieth century, Detroit was at the forefront of WTE technology as industry attempted to maximize production by eliminating wastefulness and squander. It was a city with robust democratic traditions, a strong union presence, high wages, and politics that, at least nominally, promised to overcome structural racism. At the same time, Los Angeles grew into a global powerhouse precisely because it embraced wastefulness. Neoliberalism arrived before its time and vested, monetary, interests built paradise on an ecological gamble (Davis 1998). By the end of the twentieth century the tides appear to have turned. Eventually, both Los Angeles and Detroit had to reckon with unsustainable growth and inner city urban decline. Both cities experienced a massive influx of poor non-white migrants, white flight, excessive police brutality, and massive racial violence in 1967 and 1992, respectively. Now, Los Angeles is greening the desert. In Detroit, where a swath of vacant land roughly amounts to the size of Paris (Gallagher 2013: 18), the 'desert' appears to have taken over the city.

The garbage crisis at the centre of this essay sheds some light on those reverse dynamics. Certainly, the differences between Detroit and Los Angeles remain. These are most striking with regard to political regimes and urban economies that characterize each city. For most of the twentieth century, a coalition of workers and citizens, though rife with fission and fraught with betrayal, kept the dream of social and racial equality alive in Detroit. In contrast, property ruled Los Angeles and left the dreaming to Hollywood. In Detroit, the revolt of 1967 provided the impetus for political change that culminated in the election of Mayor Coleman A. Young. The result was a near complete abandonment of the city by whites. In Los Angeles, matters were almost reversed. The Rodney King riots and natural disaster – the 1993 firestorms, the massive earthquake of 1994 and the horrendous floods the following year – led up to the end of Tom Bradley's tenure (Davis 1998: 7).

As different as these dynamics are, they play themselves out in remarkably similar fashion when it comes to garbage. In Los Angeles, the city's first and only black mayor is the author of the nation's most radical recycling programme. Remarkably enough, the greening of Los Angeles recycles the nation's profligate lifestyle and systemic wastefulness (O'Brien 2011) rather than producing sustainable alternatives. At the same time, Detroiters still see Young's incinerator as the major roadblock to a greener future – the ultimate impediment to recycling and zero waste – rather than recognizing it as a tried response to urban crisis. Captive to history and blind to the future, garbage, it seems, reproduces the system in which it is generated. The citizen-consumer of the Global North grapples with the bulky traces of overconsumption. However, our green-coloured glasses obscure that recycling – much like dumping and burning – hides the systematic production of waste and enables the continuous replacement of goods with greener alternatives. As the garbage histories of Detroit and Los Angeles illustrate, garbage practices further rather than correct for the endemic wastefulness of global capitalism.

Here and Elsewhere not only the smell of garbage is indistinguishable. As exemplified in the cases of Los Angeles and Detroit, garbage follows a similar logic – if a radically different route. In each case, garbage is systematically pushed far and further from affluent communities, effectively hiding the discontents of capitalist modernity. In each case, this development represents a continuation of previous garbage practices that reflect locally rehearsed ideals of urban citizenship. And in each case, the future is imagined as recycled past. As long as Detroit was a booming industrial city, energy consciousness and technological solutions were part of its economic model of urban efficiency. In the face of urban decline, the city thought of industrial solutions to prevent the slippage into a post-industrial future. In Los Angeles, garbage it seems has always been a private matter, occasionally erupting into the public sphere. The ensuing crises propelled homeowners' associations to ensure that the propertied are spared the public nature of waste management. In Detroit too, garbage privacy increases with affluence. While Los Angeles will transport its garbage Elsewhere – at astronomical costs – Detroit already lives there.

Notes

1 The histories of Detroit's and Los Angeles' garbage regimes are grounded in careful study of hundreds of newspaper articles and draws on select archival sources from the Bentley Historical Library at the University of Michigan.
2 Detroit Renewable Power is the official designation for the waste combustion plant with energy recovery on 5700 Russell Street, Detroit. Environmentalists and opponents talk about the incinerator instead, leaving two almost entirely separate legacies. My indiscriminate and parallel use of both designators is more than a simple refusal to pick sides in the political debates over WTE technology. Rather, in breaking with such language rules I draw attention to and call into question the moral valences of different designators for one and the same thing: Incineration is not a synonym for pollution and WTE is not always green.
3 Liddle & Dubin, PC, has filed a class action suit on behalf of the residents in the vicinity of Detroit Renewable Power. See www.ldclassaction.com/class-action/waste-incinerator-odors-lawsuit-data-sheet/ (accessed on 26 August 2015).

Part II

Excess

7 Leftover space, invisibility, and everyday life
Rooftops in Iran

Pedram Dibazar

This chapter is about rooftops in Iran as leftover spaces. Its starting point is the observation that, as a consequence of the ongoing processes of neo-liberal urban transformation, common residential rooftops in Iran are cast off as 'wasted spaces' in terms of planning and the values associated with it. The term 'leftover space' is therefore used to describe an indeterminate condition of being left out of the systems of spatial configuration and signification, which subsequently instigates exclusion from the orders of the visible and sensible. By analysing rooftop protests in Iran, this chapter argues that the Iranian residential rooftops' contours are rendered ambiguous in everyday practice, specifically in terms of visibility and systems of control. My argument is that such practices sustainably disrupt the orders of the visible by having recourse to tactics of anonymity and inconspicuousness, in ways that enhance – rather than repudiate – the conditions of indeterminacy, insignificance, and non-visibility that the rooftop fosters, precisely on the account of its leftover spatiality.

In the following, I will first outline the concept of leftover space as pertinent to the study of Iranian rooftops. Next, I will briefly explain the historic, social, and cultural bases for the application of this concept to residential rooftops in contemporary Iran, and I will explain how the proliferation of satellite dishes conflates the orders of the visible associated with leftover spaces. In the final section, I will provide an in-depth analysis of the ambiguous and confrontational trajectories of Iranian rooftops in everyday life, by focusing on the practice of shouting from rooftops at night as a form of civil protest, which is associated in Iran's recent history with the Green movement.

On leftover space

Leftover space is a contested term in urban studies, often used interchangeably with a range of definitions that denote the spatial properties of being neglected, lost, derelict, vacant, blank, slack, marginal, and void (Doron 2007b; Carmona 2010). Broadly speaking, it alludes to seemingly empty, uninhabited, or uninhabitable spaces whose form, function, boundaries, and aesthetics do not comfortably fit into the physical arrangements or conceptual frameworks of urban planning. Urban literature mostly considers the indeterminacy of such spatial conditions as an undesirable side effect of modern urban planning, caused by either negligence in the initial processes of design (space left over after planning, such as the margins of cities), failure in maintenance, programming, and after-care (space left over after use, such as old industrial sites), or inability in achieving sustainability (space left over after the living, such as wastelands). Such grey zones are thought to pose a threat not only to the appearance of a desirable city but also to the function of a cohesive society. Imprecise, ill-defined, and under-utilized, leftover spaces are commonly considered breeding places for illegal activities and dangerous behaviours.

To solve the problems posed by leftover spaces, the overall strategy developed in urban literature is the implementation of the concept of 'appropriation': conceiving creative ways to reverse the threat by reclaiming the void as a resource for carving out new concepts of public space. In this process, two antithetical processes are envisaged. Urban design and planning professions, on the one hand, aim at recuperating such forgotten spaces into the desired domains of economy and spatial order, in effect extending their managerial and ideological reach to those ill-managed sites. Processes of redevelopment and regeneration in contemporary cities are exemplary of this total planning attitude (Carmona 2010; Trancik 1986). On the other hand, the claim is frequently made that such leftover spaces open up avenues for diverse and spontaneous ways for people to make use of space in everyday life, therefore producing multiple spatialities, not necessarily in accordance with the proper orders of the space as defined by law. Advocating creative uses of space that resist given definitions of the public realm and that defy real and metaphoric boundaries of space, this second approach – illustrative of which are the postulations of 'everyday urbanism' (Chase *et al.* 1999) and 'everyday city' (Hubbard 2006) – sees in leftover spaces potential for hidden and unacknowledged counter-publics.

In other words, constant contestation over the use, and therefore definition, of space runs between the systematic processes that seek to maintain the status quo by recuperating leftover spaces – leading to more homogeneous urban environments – and the vernacular everyday practices that look for alternatives to the hegemonic order in such indeterminate settings. It is in part following this line of thought that I argue for the uncertain premises of rooftops in Iran as grounds for contestation between competing regimes of control within everyday practices. However, central to the spatial condition analysed in this paper is the perpetuation of conditions of indeterminacy in ways that defy easy appropriation and categorization into one or the other regime. As I will explain in the following, it is in exploring such a sustained condition of indeterminacy that I believe the term 'leftover' is helpful, on a conceptual level, in complicating any attempt to categorize such spaces by conventional definitions of meaning, aesthetics, or functionality.

Inherent in the notion of the leftover is, first of all, the temporality of before and after use, which purports a certain sense of waste and garbage. John Scanlan writes: 'in an unproblematic sense garbage is leftover matter. It is what remains when the good, fruitful, valuable, nourishing and useful has been taken away'

(2005: 13). Even if an object remains visibly and materially unchanged before and after use, Hird (2012) believes that its ontology changes in the course of this transition from a desirable matter to garbage. Therefore, she explains what defines things as garbage is their 'usability or worthlessness to human purposes', suggesting that 'no entity is in its essence waste, and all entities are potentially waste' (Hird 2012: 455). Following a similar line of thought, Scanlan refers to garbage as inexact and equivocal, that which defies neat definitions, and could be conceptualized as 'the remainder of such neatness'. In other words, he writes, 'the stuff of garbage' can best be defined in a metaphoric sense as 'the remainder of the symbolic order proper' (Scanlan 2005: 16–22). Consequently he writes:

> the meaning of 'waste' carries force because of the way in which it symbolizes an idea of improper use, and therefore operates within a more or less moral economy of the right, the good, the proper, their opposites and all values in between.
>
> (2005: 22)

I argue that leftover spaces should be read in ways that allow for the critical questioning of such moral economies. Over and above regarding the leftover space as a resource for potential uses, it is also possible to regard its uselessness – its defiance of the culturally constructed significations of value – as potential. In order to theorize a sustained critique of space as leftover, I claim, it is crucial to pay attention to the equivocality of meanings and values associated not only with the physical shape and materiality of space, but also with the range of activities, temporalities, and aesthetics that get attached to the processes of appropriation of it. In this chapter I analyse such intertwined spatial, social, political, and aesthetic processes that account for the residual and indefinite status of rooftops in Iran. By regarding Iranian rooftops as leftover spaces, I wish to highlight the power contained in them to question, if not totally transform, the dominant hegemony in everyday practice.

A second point considering the 'leftover' is that, by conjuring up waste and that which does not conform, it addresses issues of proximity and exposure. That which remains after the useful and valuable is exhausted is usually seen as posing a threat to the orders of the spatial and the visible precisely because of its assertive presence, detectability, visibility, and contiguity in everyday life. To administer both its inappropriateness and disclosure, the leftover therefore needs to be disposed of, disconnected from sense experience, placed elsewhere, and removed from everyday contact. Hird (2012: 455) suggests that our societies are overwhelmed by 'the desire to disgorge ourselves of waste and remove it from sight'. However, taking into account the indeterminacy of the definition of waste on the one hand, and the daily procedures of waste production and management on the other, waste is present and never totally removed from everyday contact. The physical and symbolic endurance of the residue is even more accentuated in the case of spatial leftovers, as a result of their historically embedded and contested geographies. Rather than losing touch with everyday sense experience, spatial leftovers obstinately establish contact with everyday life by providing ideal settings for a multiplicity of quotidian practices of deviation, transgression, and appropriation. The intertwinement of visibility, connectivity, and indeterminacy then poses the possibility of critique, since 'visible remainders', as Scanlan writes, 'stand as the evidence that something else is going on besides the conventional use materials and products are put to' (2005: 109).

It is because of such ambiguous positions regarding visibility and everyday contact that I find the concept of leftover space pertinent to analysing everyday practices of the rooftop in Iran. Being located above street level and disconnected from it, I argue that the rooftop's contours of visibility are in effect ambiguous and complicated in everyday practice. In particular, I will show that the subversive capacity of the rooftop in instigating counter-publics and giving voice to political dissent is predicated upon a twofold relation between visibility and invisibility, proximity and distance, and presence and absence.

Finally, the concept of the residual is instrumental to an understanding of the practices of everyday life that I pursue in this paper. To examine everyday life, as Michael Sheringham (2006) explains in his study of a range of theories and practices, is to be sensitive to the activities, aesthetics, and feelings that lag behind the dominant structures of thought and regimes of representation, and that are therefore left out of consideration in the processes of knowledge production. Most notably, Lefebvre writes: 'everyday life, in a sense residual, defined by "what is left over" after all distinct, superior, specialized, structured activities have been singled out by analysis, must be defined as a totality' (1991: 97). Similarly, Maurice Blanchot believes

> the everyday is platitude (what lags and falls back, the residual life with which our trash cans and cemeteries are filled: scrap and refuse); but this banality is also what is most important, if it brings us back to existence in its very spontaneity and as it is lived – in the moment when, lived, it escapes every speculative formulation, perhaps all coherence, all regularity.
>
> (1987: 13)

It is the liveliness of this inexorable remainder that serves as a rich and infinite source of creativity, criticality, and resistance to the ordered structures of space that seek to monopolize every aspect of modern human life. Sheringham, describing Lefebvre's theory, writes: 'the irreducible residue comprises basic human rhythms and biological needs that are not simply remainders but factors which, in surviving (and resisting), struggle against the forces that oppose appropriation' (2006: 149).

What follows from this attentiveness to the multiple implications of the residual is, as I will show in the following, an intertwined social, political, and aesthetic condition of indeterminacy in terms of the orders of the spatial, apparatuses of control, and the multifaceted ramifications of visibility in everyday life. By focusing on the positioning of satellite dishes on the rooftops and the practices of shouting from them as protest, I will argue that, despite being neglected in the processes of design and positioned out of reach and out of sight of the street, urban rooftops in Iran do not repudiate prospects of engagement with the everyday city. On the contrary, their exteriority to the orders of the spatial and the visible precisely raises possibilities for joining the everyday in ways that are disruptive of the orders of the sensible. They establish connections with residual practices of dissent and discarded voices of protest in unconventionally indeterminate, but affective, ways. The possibilities for critique that this paradox of spatial detachment and affective attachment provides are, I argue, premised upon the leftover status of such spaces. Iranian rooftops play out the power contained in the concept of the leftover space – as residual, wasted, and indeterminate – to sustainably destabilize positions taken for granted within the spatial, temporal, aesthetic, and political patterns.

Urban rooftops in Iran

The history of contemporary urban development in Iran shows precisely how the residential rooftops have been systematically cast off as leftover in design and planning. Since late 1980s, Iranian cities have been radically remodelled under the influence of the forces of speculative markets, that see in the renewal of urban centres the possibility for profit-making by vertically adding to the profitable square metres of cherished real estate (Madanipour 1998; Bayat 2010). Rather than being controlled, this process has been aided by municipalities that, disregardful of their own zoning regulations, have devised policies for selling 'building rights' as a means of maximizing their revenues. In the dense vertical cities that have emerged as a consequence of submission to the demands of the market, space is a scarcity that, in tune with the drive for maximization of profit, calls for prudence in the spatial configuration of new apartment buildings. Accordingly, spaces that do not fully contribute to square metres of saleable space – that are not readily categorized as indoors or functional – are for the most part considered as 'wasted', a squandering of the developer's investment and a dissipation of space. In this process, while in-between spaces of the old single house units such as courtyards, balconies, basements, and attics are either completely removed or reduced to the minimum in exchange for saleable square metres of indoor space, the rooftop is an unavoidable element that is held onto as necessary but treated as worthless in the processes of design and construction. Market yearnings for higher profit and architectural sensibilities for scrupulous design therefore combine to set forth new definitions of 'unnecessary' spaces.

As a result of such neo-liberal urban development schemes, common rooftops in Iranian cities are designed with little to no thought for their appearance and maintained absentmindedly over time. Resonating with this negligence is the invisibility of common rooftops for the unequipped eye on the street, which has led to the ignoring of rooftops in urban beautification policies. In short, left over as insignificant, urban rooftops have been systematically forgotten and severed from everyday contact. However, with the proliferation of the previously unthought-of satellite dishes on the rooftops, from the early 1990s onwards, rooftops have taken on a new meaning (see Figure 7.1). As the receiving of foreign TV channels through satellite dishes is regarded as undesirable by the state, on the basis of the state's lack of control over it, the previously unimportant rooftops have been unexpectedly charged with political significance. The government by and large regards the satellite technology as a 'cultural invasion by the West', a morally corruptive network that needs to be fought against. In 1995, the Iranian parliament passed a law against the importation, sale, and use of any kind of satellite equipment, legalizing their confiscation from rooftops.

However, satellite dishes have continuously resisted confiscation by the authorities, since their placement on rooftops effectively conflates the dividing lines between the legally binding concepts of the visible and hidden, public and private, and moral and immoral. By recounting disputes over the issue in the Iranian parliament in 1995, Fariba Adelkhah (1999) explains that the core threatening effect of satellite dishes was believed to arise from their visibility on the rooftops, as evident manifestation of unruliness and nonconformity to the moral values of the state. Rather than the content of the transmitted programmes, it was the display of satellite dishes on the roof that was considered to be morally incorrect as it intruded into the orders of publicness – and therefore subject to punishment. More recently, in May 2011, then deputy commander of the Iranian police, Sardaar Ahmadreza Radan, clearly stated that the police's priority in seizing the satellite dishes were the 'clearly visible' ones (*Entekhab.ir* 2011).

However, the application of the concept of visibility to satellite dishes on the rooftops is ambiguous. The accusation of intentionality in blatant public display of an unlawful behaviour is untenable since the surface of the rooftop is ordinarily unseen from the street. How, when, and to whom then are the satellite dishes visible? Although it is possible to bring the rooftop dishes into view from neighbouring rooftops, the premise upon which that visibility is assured is questionable. In particular, since the in-between state of the rooftop as a privately owned

Figure 7.1 Common urban residential rooftops in Iran (photograph by kamshots [Kamyar Adl]).

yet publicly disclosed space is posited ambiguously within the realm of the state's control, how can a vision from a private setting be used as an allegation of a public violation of the orders of the visible?

Through the intertwinement of ambiguous premises of the visibility and privacy of the rooftop, a state of uncertainty endues that poses a threat to the orders of the visible. I argue that the rooftop's implications of visibility stem from its spatial condition of ambiguity as a leftover space. Whereas the leftover status of the rooftop does not suggest any particular aesthetic regime of the visible, positioning satellite dishes adds specific meaning to its otherwise blank composition. Even though the issue of visibility is often invoked to tackle the problem of satellite dishes, what instigates rigorous reactions is the way in which, by the installation of satellite dishes, the previously insignificant rooftop gains significance as a site for illegal and immoral conduct. In other words, by adding satellite dishes, the uncertain spatial status of the rooftop is changed into one with a particular political message.

What is most compelling is that, by growing into a subject of debate and legislation in public discourse, the insignificant rooftops have gained a critical edge in questioning the cultural construction of such abstract, but legally binding, concepts as visibility and privacy. Furthermore, with the police's sporadically violent conduct and adventurous manoeuvres in seizing satellite dishes, the out-of-sight and insignificant rooftops have gained visibility in the media, exposed to the world as bearers of anti-establishment sentiments. The results of a Google search for satellite dishes in Iran show the extent to which the rooftops are rendered visible in the media as sites of seemingly unstoppable confrontation between the hegemony of the state – as manifest in the spectacle of the confiscated and destroyed dishes – and the waywardness of its citizens – detectable in the enduring presence of dishes on the rooftops. In the following, I will explore the confrontational aspects of Iranian rooftops by analysing the rooftop protests associated with the Green movement.

Rooftop protests

During the political uprisings in the aftermath of the disputed 2009 presidential elections in Iran, a number of rallies were organized on the streets by the Green movement; the first and most famous of which was a 'silent' rally, in which nearly three million people, according to some estimates, came to the streets in Tehran in silence. People's silence, although a precautionary strategy, in practice intensified the effect of their overwhelming presence, as the message of the demonstrations was to let the government see and feel the existence of people whose votes, the protesters argued, were not counted. The only signs of expression during the protests were small signs, here and there, exclaiming 'Where is my vote?' Although peaceful throughout the day, in the evening, when demonstrators were spreading out on their way back home, gunshots were fired, during which a number of civilians were killed. That initially peaceful demonstration was followed by a few less silent rallies on the streets, during which more people were killed. The uneven balance of power was already known to the demonstrators, who had opted for a silent and less provocative demonstration. However, the reaction of the regime – the extent to which it was eager to use its uneven power – was not exactly known beforehand.

After those deadly demonstrations of power by the government, the Green movement's street politics, which were effective to that point and unprecedented in post-revolutionary Iran, gradually died away. Subsequently, the main concern of the movement was to find ways to hold on, to resist complete annihilation, and to assure endurance. One of the forms in which the movement stayed relatively alive for a longer period of time, and undermined the monopoly of the authority over the public sphere, was by shouting from rooftops, which came to be known as rooftop protests (Ehsani *et al.* 2009). After nightfall, around 9 or 10 PM depending on the season, people would go up to the roof of their respective dwellings – mostly shared rooftops of apartment buildings – and shout 'Allah-o Akbar' (God is Great) and 'Marg bar Dictator' (Death to the dictator).

As a form of protest, the chanting from rooftops invites comparison with the more conventional form of street protest. It certainly purports to be a different form of expressivity in terms of space (rooftop instead of street), temporality (night instead of day), materiality (voice instead of banners and placards) and sensory faculties being invoked (sound instead of sight). Nevertheless, as I will explain in the following, rather than rejecting street politics, it effectively extends the reach of those politics to different spatial, temporal, material, and bodily functions.

The move from the street to the rooftop has a locational significance in the first place: it is a strategy of distantiation from the street. While the street is constantly policed as a result of the mobility that it offers, the rooftop maintains an autonomous geography, at least temporarily, as a result of being posited outside that system of flow. In that respect, by way of not being within the immediate reach of the police force, the move to the rooftop is a strategy to delay, if not completely deter, the direct counterattacks and brutalities of the police. In this context, the rooftop is a retreat to a 'less dangerous' position than the street, an escape to a less readily accessible space. Besides, the rooftop provides additional possibilities for escape by being in close proximity to each person's house, as it is always possible to run down and take shelter inside – given that the police is not yet prepared to fully relocate its field of action from the public to the private sphere. Therefore, the move to the communally owned rooftops of shared apartment buildings challenges the state's unconditional reaction to such demonstrations, entangles the police in legal limitations to its field of command, and charges its reactions with ambivalence and indecisiveness.

In addition to relocation, the spatiality of the rooftop addresses a different regime of visibility, as it remains mostly out of the sight of the eyes on the street. The temporality of night further positions the rooftop in a non-visible condition of darkness. As a result, the act of shouting rejects visual means of demonstrability and display by simultaneously mobilizing conditions of non-representability (in the face of the state's monopoly over such public media as

TV and the press) and non-recognizability (in the face of apparatuses of surveillance on the street). To put it differently, the invisibility of the rooftop provides a certain level of safety through sustaining conditions of anonymity. Massimo Leone describes this point succinctly:

> whereas diurnal slogans/chants of protests come from a visible source, nocturnal slogans/chants of protests come from an invisible source, protected by both the darkness of night and the position of the 'performers': thus, also those who, for various reasons, are unable to join the protests in a visible way, can do it in an invisible way (the less young, for instance).
>
> (2012: 350)

All in all, one might find a tactical gesture in the move to the rooftop, that constitutes a less dangerous way of exerting a certain level of voice and agency that is wound up intricately with everyday forms of expressivity. To start with, there are certain aspects of the rooftop protests that readily correspond to the practices of everyday life. While organizational efforts are required to sustain a single street rally on a specific day in a particular location, the shouting from the rooftop recurs with a daily rhythm at a predicted time in diverse places all around the city, and is ordinarily run as one among several daily errands with no special need for prearrangement. Besides, compared to street protests, it is inclusive of a larger range of social groups and generations. To give an example, while parents in a normative family seldom participate in street demonstrations and, dreading the prospect of the dangers involved in such rallies, would discourage the youngsters from getting involved, it is common that in the rooftop protests all members of a household participate collectively. This invitation for participation is directly connected to the conditions of anonymity that the invisibility of the rooftop provides, rendering the experience of shouting from the rooftop visually inconspicuous.

Since elusive practices of the everyday usually maintain an inconvenient relation with representational forms (Highmore 2002: 21), professional journalism has mostly failed to capture the rooftop protests visually. An exception is Pietro Masturzo's photograph of women shouting from a rooftop in Tehran, which has been widely circulated after winning the 2009 world press photo prize (see Figure 7.2). By portraying a generally neglected spatiality, this picture makes visible those ordinary people who are usually silenced, or at best misrepresented, in the media, as a result of the overexposure of certain others. Whereas in street protests women are for the most part either absent from the scene or only get highlighted in the media when their tighter and more colourful clothing attests to the image of a modern, secular, Western-styled subject, in this photo it is ordinary-looking women with casual clothes that are depicted. Furthermore, I believe that this photo is particularly affective because it depicts, by fixing in a purely visual medium, such ordinary women performing the otherwise non-visual act of 'shouting'. Moreover, to portray the act of shouting, the picture makes visible those dirty, trivial, and unimpressive scenes of the city that are

Figure 7.2 Women shouting from the rooftops (photograph by Pietro Masturzo, 2009).

customarily left out from consideration: bare walls and messy cooling systems next to the jumble of a construction site. Seen in this way, the photo is an attempt at depicting the leftovers of the governing orders of the visible; yet, as I will explain in the following, it does so by being attentive to those residual aesthetics and activities to the extent of sustaining the invisibilities inherent in them.

Peculiar to this photograph is the vantage point of the photographer, and by implication the viewer, as it seems to be taken from an elevated point, most probably from another rooftop. In this case, the photographer's move from the street to the rooftop is first of all a practical move, as a rooftop is visible only from a point higher in altitude. In addition, given the state-imposed restrictions on photographing in times of political unrest, the move from the street to the rooftop is, to an extent, forced. However, as the rooftops in question are privately owned, this is not just a matter of simple relocation on the part of the photographer. To be on the rooftop, the photographer has to gain admission by winning the trust of the inhabitants of the building, which usually works through such strategies as befriending them – in short, he has to be welcome up on the roof. One consequence of this process of relocation is that, in contrast to a street photographer, the rooftop photographer emerges as a member within the community of the specific rooftop that he enters.

The photographer is therefore transformed from a mobile specialist, ready to capture the moment while keeping his distance to the subject on the street, to one who lingers on along with a certain community, bound to the limits of the rooftop. As the protests take place at night, the immobility of the photographer is emphasized, as he is forced to use high exposure times, appropriate for photographing fixed objects. The time spent on the rooftop within the proximate and consistent community of the rooftop leads to the photographer's active and affective engagement in the scene of his photography. In a number of Masturzo's other photographs in his rooftop protests collection, moments of intimacy within this community of the rooftop are captured (see Figure 7.3).

Explaining the story behind this photograph in an interview with the *Guardian* (2010), Masturzo recounts how, after having had dinner as a guest at the house of a casual acquaintance, he accompanied his host and the rest of the guests to the roof in order to 'join the protests'. Describing the atmosphere on the rooftop as 'emotional', as people hugged and cried, he says:

> The image is blurry because I had to use a very long exposure. It was nighttime and I couldn't use a tripod or flash – the protesters were very nervous about being seen in the company of someone with a camera. It was also vital that their faces were not recognisable: in fact, it was difficult to convince them to let me take their picture at all, but I explained that no one would see who they were.
>
> I particularly like this picture because I loved that night on the rooftops. There was so much emotion.

The blurry disposition of the image is therefore not necessarily an inevitable consequence of the darkness, but particularly intended to maintain non-visibility and anonymity. In fact, a photographic mediation can violate the privacy of the

Figure 7.3 Photograph by Pietro Masturzo, from his Tehran series, 2009.

rooftop – as a privately owned space – by disclosing it to the public. Particularly opting for invisibility, the need for the visual containment of the rooftop is enhanced by substituting the professional ethics of transparency, impartiality, and objectivity with an amateurish, but no less dexterous, enthusiasm for affective engagement in the event. To be sure, as a hug between a man and a woman in Iran is incompatible with the public orders of the visible that carefully maintain the segregation of genders, the emotions contained in the photograph suggest a rather personal take suitable for private family albums. What I want to emphasize is that, as the photographer captures the practice of the rooftop protest by living it himself, he is positioned at a difficult and indeterminate point between representation and action. Masturzo's rooftop photos therefore inhabit the liminal space between the private and public, invisibility and transparency, amateurishness and professionalism.

It is following this logic that rooftop protests have been disseminated extensively on the internet through homemade videos uploaded on YouTube.[1] In such videos, acts of protest and recording merge as the people recording the event are at the same time participating in the protest by shouting on the rooftop themselves. This is strongly sensed in the videos since, given the amateur video recording equipment's ineptitude in capturing distant sounds, the clearest and loudest voice unequivocally belongs to the filmmaker – one who holds the recording device and shouts closest to the microphone. Indeed, as Leone describes, the condition of being simultaneously a 'performer' and the viewer, 'an actor of protest and a spectator of it (or, to be more precise, a listener to it)', is inherent to the rooftop protests in contrast to diurnal street rallies in which 'the crowd is a collective actor that stages a protest for the rest of the community and for the media' (2012: 351). In this 'nocturnal collective musical performance' there exists no separation between the stage and the audience. By merging the process of mediation through recording with protesting through shouting, the rooftop videos compellingly propel the viewer/listener to an affective engagement with the performance.

Crucial to the anti-representational nature of the videos of the rooftop protests is the invisibility of the rooftop that, paradoxically, negates channelling through visual media. Startlingly similar in form and content, in almost all of these videos the screen is almost always completely dark, making it difficult to discern anything except for a few sources of light in the distance. While the association of the temporality of the night with the spatiality of the rooftop – a sort of hidden time and space – renders the rooftop protests invisible, it is the voice uttered most powerfully from the top of the buildings which presents itself unreservedly to the city that is free from the noise of daytime, as well as to the viewers of the videos. Subsequently, what the films depict are the shouts, which are particularly affective by being juxtaposed to the darkness (emptiness) of the visual field.

Setrag Manoukian (2010) observes such rooftop videos in his careful analysis of the new forms of affective and experiential politics in contemporary Iran. Closely analysing a single 'video-poem' of the rooftop, he discerns a new form of politics emerging, which is premised on the interrelation between collective action – as exemplified in the video by the multiplicity of voices that shout – and individual, intimate sensations – as exemplified by the darkness of the image and the hushed voice of a woman commenting on the event. Furthermore, he detects in the particular gesture of shouting from rooftops and the exact chants of the protest a mechanism of direct referencing to – as citations and appropriations of – the same gestures and words used during the revolution of 1979. With this redeployment of the past as conveying new meanings in relation to the political landscape of the present, he believes a 'temporal disjuncture' has taken place in Iranian everyday lives. For him, following Agamben, the darkness of the rooftop video is illustrative of this disjuncture because of the intuitive courage it carries 'to look into the darkness', to grasp something beyond the restraints of chronological time. Manoukian's insightful analysis of the rooftop protests in the context of the Green movement, interestingly, parallels my reading of the rooftop videos in the use of a number of key conceptual and theoretical frameworks. However, I want to stress that – unlike Manoukian's paper, but not necessarily in contradiction to it – in this chapter I use the concept of the leftover as the framework for studying the rooftop protests. It is through the interrelation between the trashy aesthetics of the visual and sonic field of the videos, the casualness of their processes of production through everyday practices of shouting, and the leftover attributes of the space of the rooftop that I wish to analyse the subversive power contained in such practices.

As people do not use amplifying devices, the sound that is disseminated in the city during the nights of protest is unmediated, unfiltered, and uncontrolled. The shouting therefore maintains a bodily and performative utterance that suggests the most primitive and rudimentary way of demanding one's rights – shouting out loud. The unrefined character of the homemade videos supplements this condition of rudimentariness, downplaying the medium's intrinsic mediality. The way in which the texture of sound in these videos is shaped by the spatial and temporal attributes of the rooftop and the night is in contrast to what Thompson (2002: 2–3) describes as the disembodied soundscape of modern cities. In modern times, Thompson writes, with the proliferation of sound technology and amplifying devices, such as microphones and loudspeakers, a fundamental compulsion has existed to control the behaviour of sound in space, to purge out what could be regarded as the unwanted noise, and therefore to dissociate sound from its direct spatial bearings. The overall sonic experience of the modern city does not capture the reverberations of space, he continues, but rather accounts for non-reverberant, disembodied, and disjointed sounds, which have little to say about the places in which they are produced or consumed. In the modern soundscape therefore, Thompson believes, reverberations conceived as 'the lingering over time of residual sound in a space' are mostly regarded as 'noise, unnecessary and best eliminated' (2002: 2–3). The rooftop protests, however, are most affective precisely because they make sensible the reverberation of space, to the extent that one cannot definitively dissociate the shouts from them. One might say that, rather than clear shouts of protest in their singularity, the videos convey the whole space as protesting in reverberation. In short, the coarseness of the sound, unintelligibility of the image, and ingenuousness of the performance in these videos maintain a close relation with the spatial attributes of the rooftops as leftover space.

In the rooftop videos, the resonances of shouts near and far create a depth of the spatial field. By foregrounding and backgrounding sounds, an auditory idea of distance that embraces the city through the soundscape substitutes for the indiscernible flatness of the visual landscape. As a consequence, a cityscape is created that, unconventionally, is more attuned to sound than vision, making it poorly suited for the apparatuses of control as the elusiveness of sound, unlike vision, evades traceability and identification. Accordingly, as sources of the shouts are not seen in the videos, there exists no synchrony between sound and image. Michel Chion (1999) explains in relation to sound in cinema that a sound can be non-synchronous without necessarily inhabiting the imaginary off-screen. He writes,

> Consider as example the 'offscreen' voice of someone who has just left the image but continues to be there, or a man we've never seen but whom we expect to see, because we situate him in a place contiguous with the screen, in the present tense of the action.
>
> (Chion 1999: 4)

Such sounds and voices, he writes, are 'neither entirely inside nor clearly outside', instead they are 'sounds and voices that wander the surface of the screen, awaiting a place to attach to' (Chion 1999: 4). Yet, what complicates the issue in the rooftop videos is that this off-screen sound does not refer to any specific visual space, since the darkness of the image conflates a definite conception of the inside or outside of the screen.

In fact, it might be the reversal of Chion's description that is carried through in the rooftop videos: that it is a vision – an imagined vision of a person shouting – that is wandering, awaiting a sound to attach to. Therefore, the non-synchronous sound and image in the rooftop videos is conducive to the absence of direct referencing. As Chion maintains in relation to the silent cinema,

> it's not so much the *absence of voices* that the talking film came to disrupt, as the spectator's freedom to imagine them in her own way (in the same way that a filmed adaptation objectifies the features of a character in the novel).
>
> (Chion 1999: 9, emphasis in original)

Along the same lines, by recourse to absence of vision, the videos in question provide conditions for the imagination of the spectator to attach the voice of protests to an imagined vision. It is this imagination, intensified by the resonances of sound through the night, which is most affective and disruptive of the regimes of the sensible. It is this dissociation of the embodied voices from the vision that produces an ever-present spectral sense of hovering over the landscape. As a consequence of the absence of the vision, the vigorous presence of embodied voices transpires a presence that is emphatically felt, if not exactly seen.

Such expressive audial presence, predicated upon visual abstinence, is different from Amir-Ebrahimi's (2006) conceptualization of the strategies of 'absent presence' in Iranian society, which indicates that in order to entertain a 'more extensive presence in the public and often masculine spaces of the city', individual particularities and bodily nonconformist features need to be downplayed – in effect absented. Individualities obtain overall public presence, she argues, by managing the impressions that they leave in order to be 'protected by the disciplinary monotony' imposed on them (Amir-Ebrahimi 2006: 459). What follows are ghostly ways of being present in everyday life that are not seen or felt. Although the rooftop protests nurture conditions of spectral invisibility and anonymity, they do not insinuate such an absence, since the interrelation of the spatial, temporal, audial, and performative aspects of the act of shouting from the rooftop is particularly expressive of protest as discontent, resistance, and confrontation, and is impressive since it breaks the monopoly of the state over the public sphere by compellingly challenging the orders of the sensible by audible means.

Indeed, 'impression management', as James C. Scott argues, has always been one of the key survival skills of subordinate groups in power-laden situations (1990: 4). Yet, such tactical control over the impression that one leaves – which might lead at times to rigorous limitations that would make the person seem absent on the basis of the deprivation of her individual expressivities – is a delicate undertaking in the course of the practices of everyday life in ways that are not completely devoid of moments of confrontation, defiance, and critique. To understand those personal tactics of affect control, Scott conceptualizes the notion of 'hidden transcripts', in opposition to 'public transcripts'. He writes

> If subordinate discourse in the presence of the dominant is a public transcript, I shall use the term *hidden transcript* to characterize discourse that takes place 'offstage,' beyond direct observation by powerholders. The hidden transcript is thus derivative in the sense that it consists of those offstage speeches, gestures, and practices that confirm, contradict, or inflect what appears in the public transcript.
>
> (Scott 1990: 4, emphasis in original)

To the extent that the rooftop protests take place offstage and off-screen, employing diverse strategies of non-visibility, they pertain to such a hidden transcript as a vehicle through which one could 'insinuate a critique of power while hiding behind anonymity or behind innocuous understandings of their conduct' (Scott 1990: xiii). However, as the rooftop protests conflate the status of the stage and backstage both in the real act of shouting and in the distributed videos, they encroach upon the public transcript by influencing the soundscape. In that regard, the rooftop protests do not stay put on the side of the hidden, or the absent, but provide that liminal condition in which the hidden transcript meets the public one, affecting the contours of both, and sustaining an in-between space of nameless potentiality.

Finally, since this liminality is conditioned on visual, audial, and perceptive constituents, I want to turn to Jacques Rancière's definition of an aesthetic act as 'configurations of experience that create new modes of sense perception and induce novel forms of political subjectivity' (2004: 9). Rancière describes aesthetic regimes as,

> the system of a priori forms determining what presents itself to sense experience. It is a delimitation of spaces and times, of the visible and the invisible, of speech and noise, that simultaneously determines the place and the stakes of politics as a form experience. Politics revolves around what is seen and what can be said about it, around who has the ability to see and the talent to speak, around the properties of spaces and the possibilities of time.
>
> (Rancière 2004: 13)

It is by disturbing such orders of the visible, by introducing novel forms of sense experience to the partitions of time and space, that the rooftop protests provide a specific 'aesthetic-political field of possibility'. The political significance and potency of the rooftop protests, thus, does not simply emanate from politically charged words that are vehemently spoken against the power in an act of protest. Rather, it is the interruption of the distributive systems of the sensible that the rooftop protests substantiate – that which Rancière considers to be the essence of politics.

Conclusion

I want to conclude by reiterating that what sustains this potentiality for politics is the way the rooftop protests constitute the everyday. Central to this argument is the resonance between the insignificance of the spatiality of the rooftop, as out of reach and out of sight, and the anonymity, inconspicuousness, and unmarkedness of the practices of shouting from rooftops at night. Contributing to a different regime of aesthetics, rooftop protests capture a liminal space of unremittingly resilient and oppositional potentiality for radical public presence by being appreciative of the residual elements contained in their disposition in terms of space, aesthetics, and everyday practices.

Note

1 See, for instance, www.mightierthan.com/2009/07/rooftop (accessed 26 August 2015).

8 Writing rubbish about Naples

The global media, post-politics, and the garbage crisis of an (extra-)ordinary city

Nick Dines

Introduction

In 2007, reports about an urban refuse crisis in the southern Italian city of Naples began to circumnavigate the globe. Over the next four years the international media frequently featured stories and images of a city submerged beneath mounds of uncollected garbage, and overrun by popular and at times violent protests. What was actually the climax to a long-running waste emergency that had been officially declared in 1994, the escalating drama was compounded by the self-realization that the city and its trash had now acquired international newsworthiness. For Peter Popham in *The Independent*, the planetary attention appeared to confirm an apocalyptic diagnosis: 'the fear is that the Naples disease, which has put its rubbish-clogged streets on the world's news bulletins and newspapers day after day, is beyond cure' (Popham 2008). Meanwhile, Italy's mainstream press and politicians were concerned that foreign interest in the city's trash had spiralled out of control, reaching national audiences that were not usually included in their North-Atlantic-centric vision of a 'globalized' world. Hence, Goffredo Locatelli lamented in *La Repubblica* that 'the ugly images of Naples and its surroundings have ended up on the front pages of the world's newspapers, from Peking to New York, Tel Aviv to Montevideo' (Locatelli 2008).

The key issue here is not that the global media exaggerated the gravity of the problem but that its coverage was universally characterized by serious misconceptions and omissions. Analysts in Italy have detailed how the crisis was the upshot of corporate malpractice and institutional complicity that prioritized a money-spinning but unworkable plan centred on incineration, and that organized crime, although intent on exploiting the situation, was not a determining factor (Gribaudi 2008; Rabitti 2008). In stark contrast, the international media – from the world's leading liberal dailies such as *The Guardian*, *The New York Times*, *El País*, *Le Monde*, and *Süddeutsche Zeitung* to the press of emerging nations such as Brazil, China, and India – invariably conflated the breakdown of the city's refuse disposal system with the separate issue of toxic waste dumping and, by inverting cause and effect, pointed the blame at the city's criminal organization, the Camorra. Similarly, newspapers tended to interpret local people's participation in anti-landfill and anti-incinerator protests as under the sway of organized crime or as incontrovertible expressions of Nimbyism that blocked a viable solution to the crisis, while environmental activists' arguments for the causes of the crisis and their counterproposals went unreported. It is important to note that although the Italian mainstream press was generally more attuned to the institutional and legal contexts of the garbage crisis, its reporting of events was likewise marked by inaccuracies and – especially in the case of the protests – by the crude stereotyping of Neapolitans. However, the fact that local and national media coverage was itself the focus of widespread public criticism (Gribaudi 2008; Petrillo 2009) was a sign that, at least in Italy, the crisis was a contested issue.

The aim of this chapter is not simply to highlight the international media's failure to accurately tackle Naples's refuse crisis. Such an exercise would certainly be important insofar as the predominant global storyline has not been subjected to serious scrutiny (although see Dines 2013 for the British press) and because the same false claims and omissions have sometimes filtered into scholarly discussions of the affair.[1] Rather, the more pressing and interesting task here is to attempt to understand why a particular version of events became so dominant and uncontested *at a planetary level*. Dissecting the coverage of the Naples refuse crisis raises broader questions about the visibility and invisibility of garbage in the global arena. It also compels us to engage with the politics of waste governance. Reformulating Mary Douglas's famous refrain that dirt is 'matter out of place', we need to appreciate how the *rightful* place of garbage is not fixed, but rather the site of agonistic dispute.

The first part of the chapter briefly charts the evolution of the refuse crisis and the concomitant rise of an environmental movement that challenged urban waste governance, before proceeding to illustrate the news frames commonly used to report events in Naples during the period. The second part of the chapter explores three interconnected reasons that are seen to have shaped representations of the refuse crisis. First, the coverage is considered a corollary of shifting global media flows and, more specifically, of the ways in which foreign news is constructed. Second, the media's particular depoliticized take on the garbage crisis can be understood to reflect what geographer Erik Swyngedouw (2010) has termed a 'post-political environmental consensus' whereby systemic disputes are foreclosed in favour of techno-managerial solutions. Such a vision seeks a clear-cut enemy that can offer a ready-made explanation (in this case organized crime) that circumvents the complexities of the issues at stake. Third, it is argued that the reputation of Naples as an exceptional city on the margins of Europe specifically shapes the ways in which notions of crisis, violence, politics, and garbage are perceived by the rest of the world.

By combining these three factors – the nature of news production, the idea of post-politics, and the contingency of place – it is possible to deduce why international news on the garbage crisis was persistently inaccurate and why important lessons were overlooked. Developing upon postcolonial urban theory, the chapter ends by suggesting that a possible route out of this conundrum is to insist on the need for Naples to be understood as an 'ordinary city' (Robinson 2006). Such a perspective, it is argued, can encourage a repoliticization of interpretations of the city's waste crisis, while at the same time making us alert to the techniques through which Naples and its garbage continue to be framed on the global stage.

The collapse of an urban waste system and the rise of popular dissent

The refuse crisis in Naples and the region of Campania has a complex history that stems from the Italian government's decision in 1994 to move away from a dependency on landfills and towards industrial waste treatment and recycling. By declaring the waste emergency, the government delegated power to a special authority, the 'Government Commissionership for the Refuse Emergency in Campania', which was assigned the responsibility of implementing a regional waste management plan. This never became fully operative, largely due to a badly designed tender drawn up by the president of the regional government that was devoted to market solutions and the construction of incinerators (Pasotti 2010). The contract was awarded to the FIBE consortium controlled by Impregilo, Italy's largest civil engineering company, on the basis that it could offer a cheaper service and quicker start-up, despite the lack of a logistically or technologically viable plan (Gribaudi 2008). Impregilo was primarily interested in the lucrative prospect of incinerating waste because it would receive state subsidies for alternative energy production, and because the final contract was tweaked in its favour (countersigned by the regional president Antonio Bassolino in 2000), omitting improvements to recycling and the provision of back-up dumps. Attention instead concentrated on transforming as much rubbish as possible into refuse-derived fuel (RDF), which, thanks to Impregilo's negotiated exemption from the legal obligation of burning the garbage in neighbouring regions prior to the completion of incinerators, resulted in millions of so-called 'eco-bales' being stockpiled on agricultural land (Rabitti 2008). Furthermore, many eco-bales contained undifferentiated garbage, which made them theoretically unusable for incineration. The inability of the RDF plants to process the constant influx of household waste led to intermittent backlogs and, in turn, the breakdown of the disposal system and the search for emergency landfill space.

The rationale for declaring an emergency in 1994 was not only that landfills had become saturated, but that many were also under the control of the Camorra. Moreover, the Camorra's territorial power in certain parts of Naples and Caserta provinces coupled with the collusion of industrialists and local politicians enabled members of the organization to branch out into the illegal trafficking and dumping of hazardous waste. However, neither the development of the new waste management plan nor the periodic collapse of the disposal system was the outcome of Camorra interference. On the contrary, the Camorra's capacity to exploit the failures of waste governance – in particular through its control of haulage companies that were inevitably called upon during emergencies – reflected 'the historic incapacity of public administrations and institutions in Campania to control the consequences of its actions' (Gribaudi 2008: 32).

When the refuse crisis deteriorated at the end of 2007, with the suspension of garbage collections and mass protests following attempts to reopen an old land-fill in the low-income suburb of Pianura, there already existed an embryonic environmental movement that denounced the inefficient, noxious, and undemocratic aspects of Campania's refuse plan. Organized opposition to waste governance had first materialized in 2003, during mobilization against the construction of Impregilo's first planned mega-incinerator outside Acerra (15 kilometres north of Naples), which culminated in a 30,000-strong demonstration through the town in 2004. Here Impregilo had taken the decision to locate an incinerator on contaminated land in a zone notoriously dubbed the 'Triangle of Death' (Senior and Mazza 2004), due to the above-national-average cancer rates caused by illegal dumping and previous chemical industrial activity. The proposed incinerator was not only considered an impracticable solution to the region's waste crisis but also a damning indictment to a community that had been promised a new hospital on the reclaimed land.

The environmental movement that emerged during the period was socially and politically heterogeneous and involved local residents, anti-capitalist activists, mainstream environmental organizations, and groups of public intellectuals as well as dissident experts who had previously worked on the government's scheme (Musella 2008; Petrillo 2009; Armiero and D'Alisa 2012; Capone 2013). While much of its direct action concentrated on resisting the building of landfills and incinerators, the movement elaborated a comprehensive critique of the waste system that informed on health and technical matters, such as the unsuitability of selected sites, and exposed corporate and institutional responsibilities. It also drew attention to the political implications of waste governance, in particular to the continual use of emergency legislation that excluded citizens from decision-making and bypassed ordinary procedures, such as environmental impact assessments. At the same time, the movement developed a range of alternative strategies, from experimenting with door-to-door collection in order to increase recycling rates to promoting zero waste as a long-term democratic and sustainable solution. Moreover, it often moved beyond the specific confines of waste politics to engage with broader debates around environmental sustainability, neo-liberal governmentality, and democratic participation in land use decisions, which led to alliances with territorial campaigns in other parts of the country, such as the NoTAV struggle against a high-speed railway in Val di Susa in Piedmont.

The global media reports the Naples garbage crisis

Despite the high degree of complexity and controversy, the global media paid extremely little attention to the institutional, economic, political, and legal contexts of the waste crisis. The core news value of the affair was the fact that household waste was exceptionally out of place on the streets of a southern European city, rather than concealed in bins, landfills, or processing plants. This immediately connected to a set of equally dramatic themes such as violence, toxicity, and disease, which disrupted the mundane routines of urban life.

Explanations for the crisis were usually cursory and pointed to a few apparently indisputable facts: the dumps were full, the incinerators were unfinished and local protests were preventing progress. These were sometimes accompanied by vague, unsubstantiated allegations about state inefficiency or political intrigue. The Camorra was almost always identified as the source of the problem, or at least as the principal aggravating factor. For instance, Elisabetta Piqué in the Argentinian newspaper *La Nación* noted 'since the garbage crisis erupted last January, due to a lack of dumps, political mismanagement and under the shadow of the hand of the Camorra, tourism has fallen sharply in this beautiful part of southern Italy' (Piqué 2008). John Hooper in *The Guardian* was more explicit: 'This is not just about technology or logistics. It is not just about levels of garbage segregation or types of waste incinerator. It is also about organized crime in the form of Camorra, Campania's mafia' (Hooper 2008).

The urban garbage was also constantly tangled up with the topic of toxic waste dumping in the countryside outside Naples, without a clear reason provided as to why these might be connected. *Le Monde* journalist Jean-Jacques Bozonnet wrote

> For thirty years, the waste has been the business of the Camorra. The Neapolitan mafia manages hundreds of illegal landfills. But urban waste is merely the tip of a huge emergent market. The area around Caserta is full of industrial and often toxic waste, which is imported from across the peninsula and even abroad.
>
> (Bozonnet 2007)

The simple fact that hazardous waste dumping was largely controlled by organized crime appeared to confirm the Camorra's entrepreneurial prowess in all waste matters, but the conflation also suggested that any waste that was not properly disposed of (and therefore out of place) was considered part of the same problem.

Despite the steady interest in the waste crisis between 2007 and 2011, news articles were rarely followed up by in-depth investigations and never took into account the numerous critical analyses readily available in Italian. Instead, apocalyptic images of a city 'beyond cure' were hinged on a general sense of chaos, official government statements, and the alarming results of medical research. Meanwhile, key elements necessary to assemble an accurate account of the crisis – the Impregilo corporation which won the waste contract, the CIP6 government subsidies that were channelled towards incineration, and the millions of stockpiled RDF eco-bales – barely received more than a passing mention. (The lack of coverage of Impregilo was particularly striking given that the corporation's involvement in controversial engineering projects such as the

Kárahnjúkar dam in Iceland and the proposed mega-bridge between Sicily and Calabria had recently been a focus of global media attention.) As such, there was next to no discussion of the economic interests underpinning waste management or the flaws and irregularities in the tendering process.

One could commiserate with journalists having to make quick sense of something as prosaic as urban waste under such intense circumstances. And yet, back in March 2007, a delegation of foreign correspondents had been briefed about the impending crisis by a group of experts during guided visits to landfills, an RDF processing plant, eco-bale storage sites, and the incinerator construction site in Acerra. The tour was organized by the Assemblies of Naples, a group of intellectuals and scientists who since 2004 had campaigned for more accountable and sustainable waste management, had participated in protests against landfills, and had publicly denounced the illegal dumping of industrial waste (Capone 2013). A short film of the day's events produced for an Italian environmental affairs television programme portrays the tour guides disclosing legal, political, and medical information about the crisis, while the journalists offer their reactions to camera.[2] A Dutch reporter claims that the smouldering roadside tip needs to be seen to be believed, while the Israeli president of the Foreign Press Association in Rome declares: 'What we have seen today is beyond the stretch of the imagination and anything that I have read. It is truly terrible, something out of science fiction'. Inadvertently, the film reveals two very different attitudes about the trash crisis: on the one hand, the indignant voices of local activists who seek to encourage a broader understanding of the threats to public health and the environment; on the other the emotional responses of journalists who appear more attracted to the spectacle of fetid fumes and deformed lambs.

Of the 11 correspondents who participated in the tour, only three briefly recounted the experience in reports, that were overshadowed by other stories (the Camorra and its various illicit businesses) and by opposite perspectives, such as that of Guido Bertolaso, special commissioner for waste management (who had in fact been rebuked during the tour for overriding regional environmental legislation after reopening an old dump near Caserta). By the time the crisis reached a climax at the beginning of 2008, all trace of the insights into the 'failed management of the urban waste cycle' (Assise di Napoli 2007), to use the words contained in the tour's press release, had practically disappeared from the same journalists' coverage. Instead memories of the smouldering illegal dumps, which had only been one aspect of the itinerary, were recalled to enhance the drama that was now on the streets of Naples. The unfinished Acerra incinerator, which the tour guides had presented as a symbol of waste mismanagement and undemocratic decision-making back in March 2007, was now identified as the only solution to the turmoil. As Swedish correspondent Kristina Kappelin exclaimed, 'right now it is a tiny light in the darkness of a fourteen-year environmental catastrophe' (Kappelin 2008).

Besides the garbage itself, the media took a keen interest in the anti-landfill protests. Indeed, the garbage conflicts in and around Naples probably received the most sustained coverage out of all the uprisings that have occurred in

Southern Europe over the last decade. However, unlike the Indignados and anti-austerity movements in Spain and Greece, Neapolitan trash activism never contributed to a global discourse about an insurrectionary Mediterranean. On the contrary, the political nature of these protests was almost entirely overlooked.

Both the international media and the Italian mainstream press employed a similar set of overlapping rhetorical frames. First, the protests were usually connected in some way with the Camorra, either because they were seen to be manoeuvred by the criminal organization or because protesters were themselves concerned about the Camorra's influence upon waste management. For instance, John Hooper in *The Guardian*, commenting on the protests in Pianura, asserted: 'Demonstrators are frequently egged on by the Camorra as a way of keeping its grip on the waste cycle' (Hooper 2008). Second, reports tended to emphasize acts of gratuitous violence and uncivil behaviour. The fact that people set fire to public buses and hacked down trees to make barricades was evidence of their indifference to environmental matters. For example, *Der Spiegel* correspondent Alexander Smoltczyk compared the violence in Naples to the 'riots' in the Paris banlieue, declaring: 'One thing is certain: there are enough young men in the suburbs who will take any opportunity to challenge the State, whether it's in the football stadium, on tax returns or in front of the garbage dump' (Smoltczyk 2008). Third, while the media often acknowledged local residents' concerns about the public health impact of unwanted waste facilities, their motives for protest were deemed to be self-interested and familistic rather than being directed at collectively resolving the crisis. Bettina Gabbe wrote in *Hamburger Abendblatt*: 'Residents want to see an end to the crisis but they don't want landfills or incinerators on their doorsteps, just like the inhabitants of all other towns in Campania' (Gabbe 2007). Finally, by standing in the way of realistic, scientifically proven solutions protesters were by implication irrational and ignorant. Thus, Ian Fisher of *The New York Times* commented, 'in theory, a permanent solution is not difficult, and has been proposed by an emergency commission: greater recycling and the opening of several incinerators and new dumping sites in Naples and the neighboring provinces. But ... local people protest loudly' (Fisher 2007).

Criminalized, lumpen, Nimbyist, and irrational: the chances of the media extracting anything political from the trash conflicts were at best minimal. In 2011, *The Guardian* took the unprecedented step of identifying positive examples of people 'taking their city's trash problem into their own hands' (Kaye 2011). However, these grass-roots initiatives were presented as acts of responsible citizenship and social innovation rather than as constitutive of a broader political critique or environmental movement. Moreover, they remained framed against the customary exegesis of the garbage crisis that spoke of 'mafia involvement, government incompetence, and bureaucratic bungling' (Kaye 2011).

The garbage crisis as a foreign news product

So how does one explain the pervasive misrepresentation of the Naples garbage crisis? Clearly, there was no conspiracy at work: facts were not wilfully kept from audiences and the crisis was not perceived by the media to be unduly controversial. And while there were numerous instances of mediocre journalism, the case cannot simply be reduced to a question of professional deontology.

At an immediate level, the multiplication of the same incomplete and inaccurate accounts would appear to attest to the homogenization of global news flows. The economic restructuring of media networks and the dominance of international news corporations have resulted in the reduction of foreign correspondents and a greater dependency on a small number of news agencies (Clausen 2003), which together are seen to have had a negative impact upon the quality of news reporting and the proliferation of what industry insiders term 'churnalism': unoriginal and uniform news information (Davies 2008). Many newspapers from emerging nations that reported the Naples garbage crisis depended heavily on copy from international wire services. The Brazilian daily *Folha de S. Paulo* drew its news directly from the Spanish agency Efe, the Italian agency ANSA and BBC Brasil, while India's *The Hindu* used Associated Press or reproduced articles from *The Guardian*. However, this does not explain why the Italian-based foreign correspondents who participated on the 2007 tour wrote reports that were not substantially different from those produced by news agencies. Moreover, there were some wire stories, such as the arrests of Impregilo employees in May 2008 on charges of illegal waste disposal and fraud against the Italian state, that were not picked up by many newspapers.

Rather than dwelling on the consequences of the globalization of media flows, it would be more productive to think about how foreign news is selected, contextualized, and presented and how, in the case of the Naples refuse crisis, these mechanics operated in a very similar way across national media boundaries. There has long been a general assumption that audiences around the world are less interested in foreign news, unless the topic is an international armed conflict or diplomatic crisis that involves the home nation (Allen and Hamilton 2010). Therefore, the criteria that structure the selection of events are considered to be more acute than in other fields (Galtung and Ruge 1965; Harcup and O'Neill 2001). Key news values include *threshold* (the extent or intensity of an event), *unambiguity* (the clearer an event, the more likely it is to become news), *meaningfulness* (an event has some connection with the home country either through the involvement of compatriots or conformity to cultural frames of reference), *negativity, unexpectedness,* and *reference to elite people and nations.* At a cursory glance, the climax of a refuse crisis in a European city – with mounds of rubbish on the streets and people fighting the police – would appear to be sufficiently negative, intense, and clear-cut to justify attention.

Foreign correspondents, of course, do not hold a mirror up to the world but create news that will be intelligible to an assumed audience. This means translating and constructing stories within 'maps of meaning' – Italy, Naples, mafia, dirt, and so forth – that already form the basis of an audience's cultural and social knowledge (Hall *et al.* 1978: 54). So, for example, organized crime enhances the drama precisely because it fits into popular perceptions about southern Italian society. Indeed, the garbage crisis coincided with a heightened interest in the Camorra, that had arisen following a violent feud in Naples in 2004 and the international publication of Roberto Saviano's *Gomorrah* (Saviano 2008), which provided the global press with an accessible, journalism-friendly account of the Neapolitan organization and its manifold illegal activities (Dines 2013: 413–15). Maps of meaning may vary across national boundaries. In the case of the garbage crisis, such differences were perceptible in the range of tones used by journalists. For example, some of the more sober and culturally sensitive articles were to be found in *La Nación*, which may be attributable to the presence of a massive Italian diaspora in Argentina, while the British press was at times characterized by an aristocratic haughtiness that explicitly resurrected the historical memory of the Grand Tour. However, it would be spurious to suggest that these different registers constitute a 'global public sphere' (Cottle 2009: 30–32) when all the newspapers essentially peddled the same version of events.

The nature of foreign news construction therefore helps clarify why the garbage crisis became a global media phenomenon, but it does not fully explain why certain news lines, first and foremost the correlation between organized crime and uncollected garbage, became so dominant and unquestioned, nor why so many articles contained the same fundamental omissions. Why, for instance, were there no reports or images of the hundreds of acres of stockpiled eco-bales on the edge of Naples? These veritable cities of trash were just as much 'out of place' as the garbage on the city's streets. Tackling the question, however, would have meant introducing a level of complexity that would have undermined the immediacy and legibility of the rubbish crisis as a foreign news story. But more important, it would have opened up a very different discourse about urban waste.

The garbage crisis as post-political environmental consensus

The media's choreographing of the Naples garbage crisis is at the same time exemplary of what Erik Swyngedouw has termed a 'post-political environmental consensus' (Swyngedouw 2009, 2010). Drawing on discussions about the rise of post-democracy and post-politics in the work of Slavoj Žižek, Jacques Rancière and Chantal Mouffe, Swyngedouw argues that public engagement with environmental problems such as climate change has increasingly removed the space for systemic dissent and the transformative role of ecological politics, in favour of consensus formation, technocratic management, and problem-focused governance. At the same time, these issues are 'seemingly politicized' (Swyngedouw 2010: 213) by being placed high on policy and media agendas. According to Swyngedouw,

[t]he post-political environmental consensus … is one that is radically reactionary, one that forestalls the articulation of divergent, conflicting and alternative trajectories of future socio-environmental possibilities.… It holds on to a harmonious view of Nature that can be recaptured while reproducing, if not solidifying, a liberal-capitalist order for which there seems to be no alternative.

(Swyngedouw 2010: 228)

The notion of a post-political environmental consensus resonates with the global media's portrayal of the Naples garbage crisis as an ecological disaster in need of a quick fix. Rather than interrogating the inner workings of waste governance, the situation was seen to be exacerbated by the fact that the ordinary 'liberal-capitalist order' had not been able to function smoothly. Hence, journalists adopted the common refrain that incinerators and landfills offered a technologically acceptable – indeed the only – exit strategy. It is therefore no surprise that protests tended to be trivialized, stigmatized, and evacuated of political significance. Similarly, the concept of 'crisis' is typically naturalized as an out-of-the-ordinary situation rather than a set of circumstances that were instituted by a political decision and perpetuated through emergency legislation. Hence, while the global press was eager to evaluate Silvio Berlusconi's declaration in May 2008 that his new government would resolve the rubbish crisis once and for all, very little attention was paid to the concomitant passing of a decree that transformed proposed landfill facilities into sites of 'strategic national interest' and made public protest a criminal offence, thus effectively extending the state of emergency (D'Alisa and Armiero 2013).

Post-politics is particularly useful for thinking about the role the media assigned to organized crime. The Camorra not only dramatically raises the newsworthiness of Neapolitan trash: it also operates to obfuscate the systemic issues of waste management and, by doing so, distracts from state and corporate responsibilities. At the same time, it renders the situation 'seemingly political' by making vague accusations of collusion between politicians and criminals, and invoking institutions to remedial action. As the principal accessory to the media's news line, the Camorra is reified as an already existing reality rather than something that is continually reconstituted by a set of political, economic, and social processes. Just as CO_2 'becomes the fetishized stand-in for the totality of climate change calamities', so the Camorra functions as a new 'super determinant' (Swyngedouw 2010: 222) of Naples' fate. We are rarely provided with clear, consequential reasons as to why the Camorra should control the municipal waste industry, but simply by its mere mention we are made to believe that this must be the case. One thing is clear: organized crime needs to be expelled from the urban waste cycle in order to make way for a legitimate techno-managerial solution.

The global media's coverage of the rubbish crisis therefore appears to corroborate a distinctly post-political vision. It unwittingly ends up endorsing the government's waste management plan that was at the root of the crisis. The vision contrasted starkly with what took place on the ground in Naples, where the very constituents of a post-political condition – technocratic management and consensus formation – became the subject of public dissent and political action. Indeed, it was the capacity of the movement to disrupt commonsensical arguments about waste governance that underpinned its broader reach, such as its direct influence upon the election of a new city administration in 2011, which subsequently blocked the building of a proposed incinerator in Naples, committed itself to increased recycling, and signed the city up to the Zero Waste

Charter. Nothing of the sort was ever reported by the international press: it was more interested to declare a final 'end' to the whole debacle than to follow the fluid and messy political dynamics that both the crisis and the movement engendered.

One could contend that the idea of post-politics is overstated and that it does not capture anything particularly novel (McCarthy 2013). The mainstream media has long and frequently accorded priority to the definitions of those in power, while disregarding or misinterpreting the radical demands of political movements. However, what is evident about the case of the Naples trash crisis is the absence of competing stories. Even smaller progressive newspapers that were particularly sensitive to ecological issues, such as *TAZ* (Germany) and *Liberation* (France), still pitched their reports in line with the dominant discourse about the causal role of organized crime. More important, the media's post-political vision of garbage needs to be recognized as a *situated* gaze. In other words, newspapers covered waste politics in very different ways in different parts of the world. For instance, during the same period as the Naples garbage crisis, *The Guardian* interpreted the protest against an incinerator in China as an encouraging sign of grass-roots democracy that offered respite to Western fears about the 'country's breakneck economic development [and] environmental degradation' (Watts 2009). While the same newspaper never discussed alternative strategies in Naples, it treated zero waste experiments in Asia with interest, whether as examples of good practice – 'Japanese village's strict recycling regime looks to a future free of incinerators and landfill' (McCurry 2008) – or in providing hope in the face of ecological Armageddon: 'dream of a "zero waste" Goa [...] where piles of waste are causing disease to spread' (Vaz 2009). These examples suggest that what also matters for the global media is the *place* of *out-of-place* waste.

The garbage crisis as Neapolitan exceptionalism

A fundamental reason for the extended global coverage of the Naples garbage crisis was that this occurred in a technologically advanced Western nation. But it also unfolded in a city that was widely seen to possess a particular relationship with urban modernity. From travellers to journalists, philosophers to sociologists, Naples has historically been represented as an exceptional city. This was Europe's quintessential pre-industrial metropolis: one of the continent's most populous settlements during the seventeenth century, in the late modern era it was as famous for its architectural monuments, cultural traditions, and stunning natural setting, as it was renowned for its unparalleled social and economic problems and apparent dearth of law and order. According to the classic Marxist worldview, Naples was also the archetypal lumpen city: indeed it is the *lazzarone* – the classic Neapolitan lower-class rogue – that sits at the heart of Marx's famous passage in *The Eighteenth Brumaire of Louis Bonaparte* where he lists with unbridled contempt the assorted components of the lumpenproletariat (Marx 1984). During the post-war period, social scientists argued that the presence of an impoverished mass and the persistence of *Gemeinschaft*-like relations distinguished Naples from the typical European city, where a modern class structure and division of labour had emerged (Allum 1973). Hence waste in Naples was not only associated with the garbage that was seen to proliferate in its overcrowded alleyways, but also with a social detritus that was implicated, according to some (e.g. Sales 1993), in the emergence of an unproductive, parasitic, and deviant city.[3]

While the image of Naples as a premodern exception has been subject to sustained critique by scholars in recent decades (Dines 2012a: 1–7, 59–67), it continues to haunt media interpretations of the city. Journalistic descriptions routinely point to an inextricable mix of incongruous traits; to a city that is both magnificent and dangerous, dirty, and exhilarating (Dines 2013: 412–13). Over the last 40 years, besides a few blips (football success in the 1980s, urban regeneration in the 1990s), hard news on Naples has been locked into a cycle of disasters and crises: cholera, earthquakes, sinkholes, mass unemployment, and Camorra wars. This has placed the city's perceived ambivalence and extraordinariness within a decidedly negative frame. For instance, shortly after the cholera outbreak in the city in 1973, *Der Spiegel* exclaimed: 'the long-held romantic image of Naples as the city of the mandolin and popular song has been obscured by the dreary facades of tenement buildings and mountains of rubbish. Corruption and Bourbon frippery paralyze the local administration' (*Der Spiegel* 1973).

The indelible legacy of past adversities, coupled with the city's association with dirt and disorder, impinged upon the reporting of the garbage crisis. Some journalists were unrestrained in expressing their disgust. Rosie DiManno of the *Toronto Star* spoke of 'a civilized but colossally corrupt city' that was 'staggeringly foul [...] even minus the rubbish snowdrifts' (DiManno 2011). In pantomime-like manner, British journalists reeled out the Grand Tour saying 'See Naples and Die', sometimes proffering a new appendix such as 'of the stench' (Popham 2008), which worked to reiterate the pre-existing reputation of Naples as a city of extremes. Meanwhile the anti-landfill protests revived time-worn accusations of a Neapolitan rabble capable of little more than violent jacquerie (Petrillo 2009). Even the whole population was at times arraigned for abetting the crisis. On the one occasion that *The Independent* alluded to the scandal of undifferentiated trash in the eco-bales, rather than being prompted to investigate the defects of the waste management plan, the journalist quipped that 'the people of Naples should have been educated and bullied into separating their rubbish' (Popham 2008).

If the anthropologist Ulf Hannerz is correct when he argues that a conspicuous part of the work of foreign reporters 'is not devoted to hard news and unique events but to a continuous thematization of difference itself' (Hannerz 2004: 112), then the perceived singularity of Naples lies at the foundation of misrepresentations of the garbage crisis. Naples provides an enticing canvas for the global media precisely because it coincides with the very mechanics of news production. The Camorra–garbage connection and the uncooperativeness of

Neapolitans are turned into a global vernacular through which news is produced and filtered. Given the fact that the city and its related problems are already perceived as dramatic there is no need to establish a secondary, competing inter-pretation, or to probe the complexity of the questions at stake. This means that newspapers are unlikely to introduce unfamiliar or technical issues, such as the acronyms CIP6 or RDF, which require explanation and deviate from the dominant news frame.

The garbage crisis resurrected age-old paradigms about Naples being situated beyond the confines of a 'normal' Europe, with the addition that it was now also disconnected from the mundane processes of globalization. In an article on the dilemmas of recycling in early 2008, *Guardian* journalist Mark Rice-Oxley (2008) invoked the contemporary garbage crisis in Naples to ask: 'could it happen here?' In reality, he was only interested in the end spectacle of 'rotting rubbish' and not in the underlying causes, and ultimately denied such an eventu-ality by reminding readers that 'Naples is different' because it had the Camorra. In other words, Naples reassured foreign audiences by being cast as unique: the refuse crisis was seen to be incommensurable and unrepeatable and so did not tender broader environmental warnings.

Conclusions: garbage in an ordinary city

The coverage of the Naples refuse crisis calls into question the optimistic thesis of a global imagination that is exposed 'to a growing range of competing stories, images, feelings and points of view [that] is key to ... fostering more ambiva-lence and complexity in terms of how we feel and think about the world' (Orgad 2012: 6). Conversely, it cannot simply be taken as evidence of the homogeniza-tion of news and the silencing of alternative voices that has been attributed to be the direct consequence of a corporate- and Western-dominated global media industry (Herman and McChesney 1997). Rather, as this chapter has argued, the eviscerated and misconceived accounts of the crisis need to be understood as the upshot of three interconnected factors: foreign news production, post-political readings of waste and organized crime, and embedded perceptions of Neapolitan exceptionalism.

Global interest in Neapolitan garbage waned after 2011, as the excessive piles of street-side garbage receded. Nevertheless, the crisis has added another layer to the palimpsest of representations about Naples. The city now functions as a global trope for waste catastrophe. This is epitomized in an article that appeared in *China Daily* in 2013, in which concerns are expressed about Hong Kong's delayed conversion to a waste-to-energy system. The ghost of Naples's recent past is duly summoned:

> Since the government has for so long 'fiddled' while nothing has burned, should it admit defeat and consider turning the problem over to the private sector before Hong Kong becomes the 'Naples of Asia'? In 2007, the land-fills of Naples were so badly overflowing with garbage and other rotting

refuse that municipal workers would not pick up any more rubbish from the streets to add to the filthy mess.... Today, Naples is no longer one of Italy's favourite tourist destinations. The few unsuspecting visitors who arrive are soon clasping their noses and hurrying back to the railway station to get away from the stench.

(Chiping 2013)

On first sight, this extract could be considered to constitute a cultural 'counter-flow' that taunts occidental claims to environmental and technological pre-eminence. However, such a reading is immediately tempered by the fact that Naples is assumed to be nothing more than a negative signifier disconnected from any form of geopolitical influence. Again, commitment to accuracy is waived (in fact the mounds of trash had largely disappeared and tourists had returned by 2013). Naples and its garbage remain caught between the enduring visibility of the spectacle and the invisibility of alternative narratives.

In order to move beyond this impasse – which has not only characterized global media representations but has also prompted scholarly and activist misunderstandings and/or indifference about the affair – we are required, first and foremost, to interrogate the purported peculiarity of both Naples and its waste crisis. This entails a radical shift in perspective. One possible strategy is the idea of the 'ordinary city', developed by geographer Jennifer Robinson (2006), that seeks to overcome geographical bias in urban theoretical production, and hence to provincialize the ways in which cities are categorized and interpreted. All cities invent different and internally contested ways of being urban and modern. By 'bringing [them] within the same field of analysis through the idea of ordinary cities ensures that no particular city or group of cities will a-priori determine how cityness is represented' (Robinson 2006: 171). Such an approach, as I have argued elsewhere, is particularly conducive to addressing the place of Naples in relation to Europe and the world (Dines 2012b). On the one hand it functions as a mnemonic that alerts us to the techniques through which the city has been continually reproduced as deficient and anomalous, and which recently provided the world's media with an ideal setting for elaborating a post-political discourse about trash. At the same time, by being relieved of the burden of incommensurability, Naples is, like anywhere else, entitled to inform and broaden our knowledge of urban processes. If all cities are to be understood as ordinary, then Naples and its garbage crisis have many crucial lessons to offer.

Notes

1 For example, the political scientists and Naples specialists Felia and Percy Allum noted that 'public protests [were] against sites chosen for the clean, green and energy-generating incinerators' (Allum and Allum 2008: 354), which mimicked the media's positive verdict on incineration for Naples and its assertion that protesters prevented a solution to the crisis.
2 The film, broadcast on 31 March 2007 on the state channel RAI3, can be viewed here: www.youtube.com/watch?gl=IT&hl=it&v=CPbDwrWh49g. For a list of participating journalists, see the tour's press release (Assise di Napoli 2007).

3 Historically the lumpen city has sat alongside alternative imaginaries – that of an enlightened bourgeoisie since the late eighteenth century and that of an organized working class following industrialization during the twentieth century. These competing images positioned themselves in antithesis to (and thus ultimately reaffirmed) the hegemonic idea of Naples as an aberrant city. They also had negligible bearing upon foreign media representations of the city.

9 Dirt poor/filthy rich

Urban garbage from Radiant City to abstention

Pauline Goul

Reading about garbage and social inequalities, one could expect to find economic considerations together with environmental concerns. However, this is only part of what the present chapter proposes to do. In the context of global discussions of garbage as reshaping urban communities, this essay proposes to analyse, at the intersection of urbanism and garbage, the way that architect Le Corbusier conceptualized excess and waste in *Radiant City*. Establishing the fact that garbage in cities inscribes itself in a long tradition of reactions of disgust and counter-planning can help us redefine the ways in which we imagine garbage today. The chapter also looks at the language of garbage, identifying surprising paradoxes and an ambivalent fascination for the latter in the works of Le Corbusier, but also in those of Georges Bataille. The chapter then continues to deepen the understanding of garbage by considering ecological endeavours that strive to eradicate garbage altogether, thus contrasting Le Corbusier's vision with more contemporary tendencies in global cities.

The American poet A.R. Ammons begins his poetry collection *Garbage* with the line 'garbage has to be the poem of our time' (Ammons 2002 [1993]: 13). Beyond the ironic stance there is in pretending that any product of creativity could be garbage, beyond the consideration that art or literature were originally – or at least with and after Plato – considered to represent beautiful things only, what the line really means is that garbage is a poem, or, rather, that there exists a correlation between trash and language. One has to admit that, if our time needs a representative poem, garbage fits perfectly. Since a poem is made of metaphors, Ammons is surely implying that garbage is the metaphor, or even the image, of our time. And in fact, visions of garbage represent for the media the long-term disaster footage that can be used when there is no environmental catastrophe to cover. Garbage is *the* constant environmental catastrophe, playing a leading role in pollution, climate change, and spatial limitations, and yet it does so in different ways once we look at it globally. But from impoverished neighbourhoods to affluent ones, cities always stage the frustration of garbage and consumption. Even more, garbage is a poem because it has to do with language, it *does something* to language. It is the dark place where words go when they want to signify something extreme and wrong: hence the title of this chapter, 'Dirt poor/filthy rich'. Both of these are idioms that would strike anybody as particularly telling of the economic dimension of garbage: moreover, it is undeniably paradoxical that two opposites on the scale of wealth are qualified by a very similar derogatory notion. If one takes a look at the words themselves, 'dirt' seems to imply that poor people are closer to the ground, that they tend to live in areas of the world that are not as clean and as covered in protective asphalt or as punctuated by green spaces as what is considered to be normal in 'developed' countries. Furthermore, the distinction between both extremes does not necessarily correspond to the distinction between developed and underdeveloped countries, since dirt poor and filthy rich often co-reside in urban spaces that have had or still have an accelerated economic growth, regardless of their position in the North and South divide.

This juxtaposition is visible in various cities on the global scale, from New Orleans to Paris, from Nairobi to Rio. The image that expresses this contrast most significantly, in a visual oxymoron, is a photograph of two adjacent neighbourhoods of São Paolo, the *favela* of Paraisópolis and the upper middle-class high standing buildings of Morumbi, with private swimming pools on each balcony, taken by Tuca Vieira. The picture, first exhibited at the Tate Modern in London in 2007 for the 'Global Cities' exhibit, represents exactly the urban and economic inequalities caused by globalization, as they differ from one neighbourhood to the other. Such disparities are often studied in works on environmental justice and sustainability, as in Brian J. Godfrey's 2013 chapter on Rio de Janeiro, 'Urban Renewal, *Favelas* and Guanabara Bay', which also emphasizes the position of trash in the rapid urbanization common to developing countries as they strive to catch up with the global economy. Globalization, understood as the internationalization of trade and economic exchanges, can be said to cause in equal part the affluence of certain neighbourhoods and the increased poverty of others. In countries like Brazil, which are simultaneously 'developed' and 'developing', international investments remain limited to one section of the population: that of skilled, educated workers. In the process, local production struggles result in even lower wages, while the benefits of the developing export markets are concentrated in the same areas of the population, and thus of the city. Yet Tuca Vieira's picture also shows the repartition of waste in global cities: it is invisible in rich neighbourhoods, where the colour white is predominant, and where water and cleanliness are obvious and taken for granted as a right, but it is extremely evident and scattered in *favelas*, where houses are built out of recuperated material, where the dominant colours are earthy like dirt.

On the other side, 'filthy rich' is a difficult idiom to uncover: well-off individuals obviously have access to soap, live in proper neighbourhoods, with all the necessary amenities to hide away the garbage, or even to systematically get rid of it. So why would 'filthy' end up qualifying extremely rich people? It seems that the origin of the phrase comes from another stereotype, claiming that money, in particular coins, are dirty. The origin of the association can be found in Tyndale's translation of the Greek Bible, where the greedy is guilty of 'filthy lucre' (Timothy 3:3). While at the time it implied that the wealth is ill-gotten, the more recent phrase 'filthy rich' takes perverse undertones with the evolution of 'filthy'. Yet in the aftermath of the global economic crisis of 2008, one could say that 'filthy rich' has come to mean literally filthy: the visions of dog excrement and messiness in the documentary *The Queen of Versailles* (Greenfield 2012) speak for themselves. Visually, they associate the gigantic home, in fact the largest private residence in the United States of America, of Florida billionaire entrepreneur David Siegel – the queen of Versailles from the title is his wife, beauty-queen and former IBM engineer Jackie Siegel, who insisted that her house should look like Versailles – to reality television series like, the *Hoarding: Buried Alive*, where every episode, under the pretext of 'explor[ing] the psychology behind compulsive storing' – quoted from the show's website – mostly exhibits the untidy, dirty, and overflowing interiors of victims of the disorder. The documentary *Waste Land* (Walker 2010), also set in a Brazilian metropolis, reveals the divergences between what the well-off consume, and what the inhabitants of the *favelas* find through the city's trash. Among the many contrasts that can be found in the documentary, the juxtaposition of images of the Jardim Gramacho landfill and of Vik Muniz's exhibit in Rio's Museum of Modern Art (MAM) highlights the very same extended oxymoron.

It thus seems undeniable that garbage and money are intrinsically related, complicating any attempt at conceptualizing the variations in waste production and management in the modern city. Starting with an analysis of a curious obsession for waste in Le Corbusier's utilitarian plan, his *ville radieuse*, this chapter will look at the urban particularities and qualities of garbage, at its economic repartition and the environmental issues that arise out of it. Before being taken out of sight, out of mind, to the landfill, itself generally on the outskirts of or well outside the city, what place does garbage hold in the city? According to Le Corbusier, garbage takes many forms, and it can enlighten more recent conceptualizations of urban waste management. Yet nowadays, zero waste lifestyles emerge on the social media (Instagram, blogs, Twitter) with similar aesthetics and pose the question of responsibility and economical gaps in considerations of waste: are the new 'filthy rich' literally people who can afford to be close to their garbage and to analyse it so as to reduce it? Are we making sustainability in general into something only the privileged class can afford to care about?

Paradoxically, in order to understand the close relation between garbage and social inequalities anywhere, it is productive to look back at the 1930s. In *Radiant City*, where he exposes his vision for a 'green' city, Swiss-French architect Le Corbusier declares 'the problem of architecture is the basis of social equilibrium today' (Le Corbusier 1967: 19). As one of the vanguards of modern urbanism, he takes part in what could be considered as a contemporary feeling of disgust for cities, as he wonders: 'Does the big city express a fortunate or a harmful occurrence?' (Le Corbusier 1967: 35). Robert Fishman, in *Urban Utopias in the Twentieth Century*, observes such a disgust for cities in modernist planning theory: three major visionaries of urban planning, namely Ebenezer Howard, Frank Lloyd Wright, and Le Corbusier, he writes, 'hated the cities of their time with an overwhelming passion. The metropolis was the counter-image of their ideal cities, the hell that inspired their heavens' (Fishman 1977: 12). Unsurprisingly, Le Corbusier's personal disgust – expressed in the exalted and obsessive first-person narrative of *Ville radieuse*, is fundamentally dependent on the notion of garbage. Explaining the purpose of *Ville radieuse*, he denies that the book belongs to the genre of literature:

> Il exprime le martèlement de la vie présente, la croissance accélérée et vio-
> lente d'un phénomène neuf: l'urbanisme; l'explosion des malaises accu-
> mulés, l'éclatement des crises; les impasses...

> (It describes the battering of life today; the rapid and violent growth of the
> modern phenomenon of urbanism; the explosion of accumulated anxieties
> and the outbreak of hysteria; dilemmas...)
>
> (Le Corbusier 1967: 7)

In the emphasis on violence and excess, he merely dissimulates the root of all urban evil: garbage. The 'malaises', the 'crises', the 'impasses' are arguably all variations of garbage. Le Corbusier could in fact be said to have been quite obsessed with garbage. It is striking from the first pages of *Radiant City*: at first, he really is talking about *déchets*, garbage itself: 'The city? It's already an empty shell. Its product is there all right, polished and superb, clear as crystal. The fruit of culture. But look at the refuse and scum. Look at the misery, the unhappiness and stupidity' (Le Corbusier 1967: 6). In the construction of the sentence, refuse and scum directly signify misery, unhappiness, stupidity. However, Le Corbusier arranges it in such a way that it is impossible to tell which of the two is the cause, and which is the effect, leaving the question suspended. Moreover, 'scum' translates the French *scorie*, which means the industrial residue of mining. This helps to determine the definition of Le Corbusier's garbage: it is in fact not just domestic garbage, but the entire trash production of any big city – or at least that is what his choice of words suggests. His garbage is also not limited to that of Paris, since he designs plans for Stockholm, but also, most notably, for Algiers and for Chandigarh (India). The cities Le Corbusier studies and wishes to revolutionize are global, South and North, West and East. Yet his language in *Ville radieuse* notably revolves around garbage, with a choice of words that betrays his obsessive fascination with refuse and deploys a set of dependent notions. Soon after *scorie*, the refuse coming from mining, one finds *hardes usées*, literally 'used rags', a socially charged garbage, that of homeless people, that of clothes worn out until the very last thread. Furthermore, he uses the phrase in a personification of society: 'Modern society is throwing off its rags and preparing to move into a new home: the radiant city' (Le Corbusier 1967: 7).

However, the intricate network of what Le Corbusier considers garbage expands its scope significantly: on the very next page, he mentions 'antiquated junk' (Le Corbusier 1967: 8) or *antiquaille*: he isolates a whole network of perverted relationships, the central cause of which can be identified as this 'antiquated junk':

> When nature follows the course of its seasons – birth, maturity, death, spring, summer, fall, winter – and every year cleans, undoes, buries, us, the brains of the universe, we consent to live in rotting streets and houses (…) to suffer deceased rules, to be governed by institutions that have long been detached from the tree of life, moldering in the realm of dead things: our steps, our gestures, our thoughts, all dominated by antiquated junk – with the exception of our industry – right down to our incredibly grotesque kitchen equipment.
>
> (Le Corbusier 1967: 8)

Garbage has a lot of markers for him. It is everything that is unnatural about the city, and is opposed to the 'green city' Le Corbusier strives for, envisioned as what he calls 'the tree of life', which could be taken simply as a symbol of nature. It is also, as can be said of excrement, one of the great equalizers, since it seems to subsume any difference in class, social status, or richness. He uses the first-person plural 'we' to express that every city-dweller in the world is faced with garbage, with antiquated junk. Yet this is precisely when Le Corbusier's target – garbage – seems to become a hallucinated exaggeration, since one could seriously doubt that the second arrondissement of Paris, to take one example, was ever the place of 'rotting streets and houses' in the 1930s, having always been an affluent neighbourhood. Le Corbusier is placing garbage as an allegory of the impracticality and the messiness of the accelerated growth of cities at the time. Yet, what this quotation expresses is also a tipping over of the concept of garbage into the more general notion of waste, encompassing trash, dirtiness but also, with the 'antiquated junk' that seems to refer to 'kitchen equipment', mere consumption. The dominant word is therefore not as much garbage as it is the French *gaspillage*, a noun for the verb to waste – of a related etymology. In a nutshell, it involves buying too much, from unnecessary things to too many and specific utensils, but also purchasing food and letting it go to waste. It is consumption for the sake of consumption, without any further purpose. This is where Le Corbusier actually formulates a clear criticism of the contemporary society of consumption, in a quite anti-capitalist stance.

The images illustrating *Radiant City* are therefore particularly fascinating: Le Corbusier chooses his illustrations for the book quite originally: he seems to make collages of scraps of newspapers, especially free magazines and catalogues devoted to the browsing of potential purchases, such as those that are still distributed in subway stations in Paris, or in individual mailboxes. Ornamented utensils occupy most of the space of his obsession for waste. Figure 9.1 displays vases, cups, bowls, lamps, or even electrical apparatus. Moreover, he annotates these images with almost lyrical exclamations, where his fascinating relationship to consumption can be grasped most clearly:

Le gaspillage! Je ne m'indigne pas de ce que ceci s'achète. Mais je me désole de voir l'Autorité demeurer indifférente devant un tel sacrilège: le temps perdu à fabriquer ces imbécilités! Un peuple sain, conscient, fort, devrait crier: assez!

(Waste! I am not outraged because these things are bought. But I am deeply distressed to see Authority remaining indifferent in the face of such sacrilege: the time lost in manufacturing these tomfooleries! A healthy, aware, strong nation ought to say: enough!)

(Le Corbusier 1967: 94)

Could it be that Le Corbusier's problem is precisely that he attributes the responsibility for all this waste to the decision to manufacture these objects instead of in the general tendency to purchase them, since he claims that he is 'not outraged because these things are bought'? In short, he seems to be ascribe guilt to the offer rather than the demand. Or perhaps one could say that he already understands the 'soft pollution' Michel Serres conceptualizes in *Le Mal*

Le gaspillage! Je ne m'indigne pas de ce que ceci s'achète. Mais je me désole de voir l'Autorité demeurer indifférente devant un tel sacrilège : le temps perdu à fabriquer ces imbécillités ! Un peuple sain, conscient, fort, devrait crier : assez !

Figure 9.1 Examples of excess, from Le Corbusier, *Radiant City* (courtesy of the Fondation Le Corbusier, © FLC/ARS, 2014).

propre, where consumption and pollution are themselves created by the per-petual proliferation of manipulated information coming from manufacturers and CEOs: Serres in fact argues that 'the rich readily discharge waste – another case of dumping – where the very poorest live', using Paris's beltway as an example (Serres 2011: 46). The 'aggressive images, billboards, and giant lights' are reserved to the working class northern neighbourhoods, whereas the western part, where the CEOs actually live, has greenery and no advertising. Le Corbusi-er's images come from pieces of advertisement, catalogues of items supposed to provoke desire.

In fact, the process is an old one, one that Rebecca Zorach analyses in *Blood, Milk, Ink, Gold*, her study of excess and abundance in Renaissance France. She focuses on one utensil, a golden salt cellar designed by Benvenuto Cellini, and perhaps, in her words, 'the most famous artwork made in France in the sixteenth century' (Zorach 2005: 90). As part of her study of material desires and the pro-cesses of consumption that took place in what she identifies as the nascent capit-alism of early modern France, she isolates the processes that allowed this single golden salt cellar to be desired and that allowed this desire to be spread through-out the kingdom. Zorach calls it the 'textbook example of the excessive luxury' of the court, and her analysis of the phrase 'golden saltcellar' speaks to the paradox of 'filthy rich': 'The phrase itself, "golden saltcellar," seems to suggest a disjunction of form and function – specifically of valuable material and quoti-dian function' (Zorach 2005: 90). It is therefore possible to retrace the fascina-tion for that one object throughout the various graphic works of the school of Fontainebleau, namely, the movement of Italian painters who, in the first half of the sixteenth century, came to the court of Francis I in order to decorate the Fon-tainebleau castle (Figure 9.2). Ultimately, it is possible to distinguish, in the first stirrings of an emulated desire for luxury items among the nobility in sixteenth-century France, the same disregard for practicality at the benefit of exterior beauty, and the same fascination for appearance that some consumer goods in nowadays' stores reflect.

Packaged, processed objects are most commonly bought because they bear a brand name and offer an attractive exterior. If they were to be compared, as will be the case later on, with bulk goods, one would have to admit that even if they are trendy in some zero-waste circles, they are not a quarter as popular as a good old-fashioned improved image of the food you may get once the object is pre-pared and ready to eat. The French *photo non contractuelle*, the English 'photo for information', betray the manipulation behind the packaging on our common processed object of consumption. Furthermore, one simply cannot avoid notic-ing the striking resemblance between the sixteenth-century salt and pepper cellar and the 1930s catalogue of utensils, of 'antiquated junk'. The ornamented exte-rior seems to speak of the uselessness of the object, although it has gained a 'kitsch' dimension. One could even formulate the rule that the ornamented appearance that governed consumer desire in the Renaissance has been trans-ferred into brand names, illusions of practicality, and attractive packaging. Out of several usable goods, the way we are made to choose between several of them

Figure 9.2 Salt cellars in the graphic art of Early Modern France, from Rebecca Zorach, *Blood, Milk, Ink, Gold* (courtesy of Rijksmuseum Amsterdam).

is by some sense that it would be the best choice: it is thus rendered more attractive, even if more expensive, by intelligent and careful advertisement and researched marketing. In the very definition of both concepts lies an indisputable paradox: if the object is indeed needed, then it would not be necessary to communicate that need.

In an earlier illustration (Figure 9.3), Le Corbusier introduces the distinction he makes between useless consumer goods – that he associates with junk – and useful consumer goods. His conception of what falls into 'antiquated junk' thus has to do with lack of efficiency or usefulness, unsurprisingly. Indeed, his whole project is based on the rationalization of society, with influences from Fordism and Taylorism, as in the impossible 'Minimum House', unrealizable because of the omnipresent maximum, which could be understood as a form of excess, that is to say of waste.

Gaspillage demonstrates another aspect of Le Corbusier's thought, the conflict between individual desires and collective action. In fact, *gaspillage* seems to be used in cases when he speaks of individual consumption, for personal pleasures: in his new philosophy of the city, 'communal services' are supposed to eliminate waste, precisely because waste would be provoked by contrasting

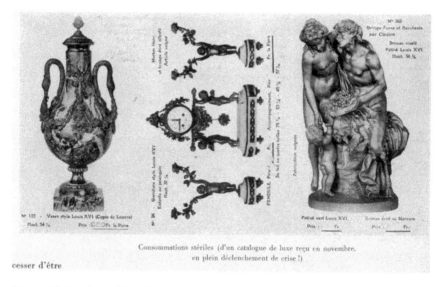

cesser d'être

Figure 9.3 'Antiquated junk', from Le Corbusier, *Radiant City* (courtesy of the Fondation Le Corbusier, © FLC/ARS, 2014).

individual desires that feed each other – from wanting what the neighbour has, to aspiring to own what the more affluent neighbour displays. The basis of consumption is indeed in one sense envy, the need to possess whatever the other is having, as Rebecca Zorach has illustrated in relation to the salt cellar. Nowadays, in crowded cities, it is increasingly difficult to distinguish the actual waste – that is to say, the garbage – from the potential waste, which can be found in the proliferating shopping centres. Consumption has become so paramount to urban living that it actually partakes in very visible ways in a city's garbage. As an example, one can consider the *catadores* of the documentary *Waste Land*, who, as they picked through the city's garbage, get acquainted with cultural objects that come from the more affluent neighbourhoods. They gather books – the president of their union, Tião, collects these books and teaches himself political theory, comparing contemporary politics to what he has read in Machiavelli's *Prince*.

Yet if Le Corbusier's obsession with waste is the intriguing point that has inspired this chapter, what pushes it further is the uncanny and paradoxical, but explainable resemblance between his aesthetic and what I would call, after a *New York Times* article entitled 'The Year Without Toilet Paper', the aesthetic of urban ecological abstention. In the following, this essay analyses two examples of urban ecological abstention: first, the *No Impact Man* project (2010), where Colin Beaver, a writer at the time, now a political ecologist, takes up the challenge of reducing his family's carbon footprint to the minimum while living in New York City, and a similar, more recent blog and book written by Bea Johnson, *Zero Waste Home*, where the mother of two children focuses her efforts on reducing waste in particular. In many ways, Le Corbusier and these recent abstainers share their will to rationalize consumption and urban living, and reduce waste, but both endeavours have irreconcilable differences: one of them is that, in contrast to Le Corbusier's *Radiant City*, which envisions a collective plan of urban waste reduction, green living is nowadays promoted as an individualistic project. In *No Impact Man*, there is even a great ambivalence between the figure of the man, Colin Beaver, and how he promotes himself as a blogger and an ecologist, and the reality of his family adopting the project as a whole – and his wife in particular getting no credit for it. *Zero Waste Home* is slightly less individualistic, in that it focuses on one home, instead of one person, but still not enough: Bea Johnson, stay-at-home mother and blogger, is at the centre of the project, pushes it forward and wins awards and fame for it, while her children and husband are sometimes interviewed for a few seconds to declare that, respectively, their new lifestyle does not bother them too much or that it truly is proven to be more economical. The frontiers of the home are much smaller than those of the Radiant City, where every attempt at limiting waste is thought of as a collective, large-scale process. Certainly, the origin of this difference is to be blamed on the lack of collective policy or institutional commitment on these crucial projects. In *Zero Waste Home*, Bea Johnson, while she manages to reduce her actual trash to a quart-sized jar a year (Johnson 2014), defines trash as becoming everyone's problem once it crosses the threshold of the home. Yet this means that the only trash that counts as reduced is the garbage that makes it to the house: Bea's whole strategy consists in refusing any access of trash into her home, a smart yet not entirely satisfying theory. The logic is clear and well thought of: by refusing a receipt at the store that has already been printed, she advocates for the fact that most customers actually do not need the printed receipt. Certainly, were everybody to adopt this behaviour, there is a possibility that stores would actually stop printing receipts altogether, or switch to online receipts. But she underwent a lot of criticism after an article about the Zero Waste Home mentioned that the family was sending back the plastic slip included in Netflix's DVD envelopes. Indeed, sending back or refusing trash does not change the fact that the receipt goes to garbage anyway, whether at the store or in her home. Her quart-sized jar omits a sizeable part of the garbage she produces out in the world, which would be reflected if it were expressed in terms of her carbon footprint. The Zero Waste Home is moreover strikingly similar to what Le Corbusier would call the Minimum House, and is perhaps even inspired by it. The minimalist design, the all-white aseptic decor, the simple Scandinavian furniture. Yet the overall aesthetic, in addition to and also in contrast to the cold, utilitarian look of Le Corbusier's drawings, does convey a distinct air of expensiveness, if not luxury. While Le Corbusier took inspiration from the cruise boat's approach to economical design and practicality, the more recent aesthetic of urban ecological abstention seems to have transferred the same concerns to the first-class cabins – in the sense that simplistic design has undeniably become an elitist trend. It therefore seems as if the care for simplistic lifestyles and designs has become a prerogative of the privileged instead of a collective concern of the majority. However helpful and crafty Johnson's tips are, however convincing her accounting and proof that her family does save money on this, her lifestyle is enabled by several conditions that are more or less independent from her: she lives in an extremely expensive area, Mill Valley, CA, in an already environmentally friendly urban space where health stores are much more developed than in most American cities, especially since a good part of her lifestyle revolves around finding groceries available in bulk. Furthermore, Johnson admits openly to having developed an obsession for waste reduction: in an article, she admits that, at the beginning, getting rid of stuff was 'very addictive' (Story 2011). Her obsession for waste reduction thus bears similarities with the behaviour of compulsive buyers and hoarders. In a recent blog post, she actually explains that her children think that she is a hoarder, because she has kept all of their waste since October 2010. She confesses another obsession in her book, while narrating how she came to adopt this lifestyle with her family: 'You might even say that I became addicted to shopping in bulk, driving far distances within the Bay Area, searching for suppliers' (Johnson 2013: 6).

In fact, if one wants to provoke some changes in terms of environmental living, the urban landscape poses particular problems; it seems easy to tend to a vegetable garden, to have a compost, when you live in the countryside or in a suburban area. In contrast, Le Corbusier's focus, being more European, and thus perhaps more urban insofar as it is less suburban, focuses on the city space, where composting would be impractical and much more of a hassle. In the documentary *No Impact Man*, Colin Beaver tends to his own compost inside his small Manhattan kitchen. Bea Johnson has an undeniably limited approach, one that makes her Zero Waste lifestyle seemingly dependent on her being a stay-at-home mother, who has at least some time on her hands, an interest for interior design, and a starting budget to invest into reducing consumption.[1] This does not change the fact that her blog and book started a trend among other families in the USA or in France, where the book has been translated. It does, however, demonstrate how, faced with the lack of public policy, ecological living becomes a predominantly socially privileged and individualistic endeavour, which could be accused of being a mere hobby. The *New York Times* article on *No Impact Man* called the various people who buy into the trend of zero waste lifestyles 'abstainers', and yet one has to notice most of them belong to a privileged social class, a paradox some could relate to what is often called 'first world problems'. Abstaining requires that the alternative, splurging and wasting, is what otherwise happens, what happens for the majority. This does risk conveying that, in order to care about reducing urban waste, one needs to be able to afford it. Is it not possible to find a more global, more universal approach to reduced consumption and waste management, one that could speak to the inhabitants of neighbourhoods or areas of the world that are actually deemed dirty, dirt-poor?

Several endeavours that are more collective strive to bridge the perceived gap between caring well-off ecologists and what is often understood as wasteful, lower-middle class individuals – one can think of the example of *Hoarding: Buried Alive*, for instance, where most of the selected 'hoarders' evoke a working-class, lower-middle-class environment, which thus becomes associated with wastefulness. While reality television plays a great part in fixating this particular stereotype, with popular shows like *Here Comes Honey Boo Boo* revealing America's uncanny fascination for what is called 'hillbilly' or 'white trash' – the term could in fact turn the dirt poor/filthy rich divide into a ternary where waste is always present in various ways and proportions. Examples of more collective endeavours thus go from habits such as gleaning in Agnes Varda's documentary, *The Gleaners and I* (2002), which notably focuses on homeless persons, Romani individuals, and struggling marginals, to recent urban movements like freeganism or dumpster diving. Moreover, public art projects such as '70 × 7 The Meal' by Lucy and Jorge Orta are not necessarily irrelevant, just because art is seen as being targeted for a privileged, educated audience. The Meal is a communicative, performative approach to sustainability and environmental awareness, relying on a potentially infinite number of benefit banquets made with local ingredients, given by local farms, to which random persons, from bankers to unemployed people, are invited. For these banquets, porcelain durable dishes are created and distributed, to be kept and reused, instead of plastic or cardboard silverware. Hence, these various artistic projects dealing with waste and the movements just mentioned all point to a similar aesthetic of recuperation, bricolage, at what we could call a rusticity – a rusticity that is reminiscent of the Radiant City. Is rustic the better way to rethink the city, instead of making it more aseptic, more hygienic, more non-human? Can the city be made more rural, for a more accessible and affordable sustainable living for all?

In fact, there is a notion that could be key to understand the issues of dirt-poor and filthy-rich in relation to their conception of environmental living, from Le Corbusier to Bea Johnson, from Varda's homeless individuals to Colin Beaver's *No Impact Man*: could it be that leisure is an essential component of any discourse on ecology? Waste in general is usually thought of as a mere by-product of normal life, but rarely is it considered as happening within a place of leisure, or at least of free time. Indeed, urban movements such as freeganism or dumpster diving are mostly led by young people, either students or unemployed, in a refusal of the regular liberal economy that seems to reject them, like the garbage it refuses to see, in its outskirts. Various scavengers, from canners in New York City, as seen in the short documentary *Redemption*, to *catadores* in Rio as they reached fame with the documentary *Waste land* survive because they use their otherwise idle time – because they are unemployed – to collect recyclables, and make unofficial jobs out of it. Leisure is indeed a critical concept in considering problems of waste; greener living in general is indeed considered more time-consuming than the common frozen meals and take-away containers. Time is thus the critical hinging point: processed food creates more garbage, but it is quicker and more efficient on a daily basis. When cities set up programmes for recycling and sorting out domestic waste in different garbage receptacles, a sizeable portion of the population complains of the time wasted. Putting trash in the garbage is the quickest and the easiest way, while reducing it takes time. Basically, if you have time to waste, you can think about how to waste less. Yet one

On bâtissait ces choses pendant que se construisaient les chemins de fer..

Figure 9.4 The extravagance of leisure, from Le Corbusier, *Radiant City* (courtesy of the Fondation Le Corbusier, © FLC/ARS, 2014).

has to admit that it does not particularly help the cause to have spokespersons that are either housewives or writers, extravagant artists, or unemployed youth. Precisely, in Le Corbusier's plan, leisure had a derogatory place, and I shall argue that it may actually be part of where the project failed or would have failed.

In *Radiant City*, an illustration of a baroque theatre comes with the caption 'Things like this were being built at the same times as our railways', obviously opposing the utility of railways to the useless and wasteful appearance of theatres. One could wonder if Le Corbusier's target is art in general, but it seems to be genuine leisure, in all its forms. Hence, later on, cultivating your own food is assumed to be impossible in the 1930s economy, when Le Corbusier deems it 'romanticism' to think that modern man can 'rest himself' from work by working his garden. This suggests that a leisure that doubles up as something useful and economical, gardening, cannot be considered leisure by Le Corbusier because it involves some form of labour. His preferred audience, the part of the population that he addresses, is the working class from the offices and the factories in cities, whose leisure time should be spend elsewhere, and completely labour-free.

So, greener living can be related back to leisure, but I would like to propose that this does not mean that its scope is limited to the prerogative of privileged individuals. In fact, the social class most concerned with leisure should be, in the work of Marxist Paul Lafargue, the proletariat. In his opinion, it is because of its unnatural passion for labour that the proletariat provokes overproduction, which in turns leads to forced leisure. He exposes these ideas in the 1880 pamphlet *The Right to be Lazy*:

₁. LES LOISIRS

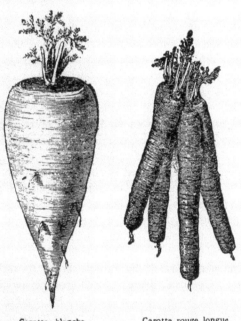

Carotte blanche   Carotte rouge longue
des Vosges   obtuse sans cœur

Faire des carottes et des navets n'est pas une
distraction. C'est un métier. On ne repose pas
l'homme moderne fatigué par le bureau ou
l'atelier, par l'épuisant travail de la terre.
                    Romantisme çà !
                (confession d'hommes sincères)

Figure 9.5 The paradox of useful leisure, from Le Corbusier, *Radiant City* (courtesy of
the Fondation Le Corbusier, © FLC/ARS, 2014).

Parce que la classe ouvrière, avec sa bonne foi simpliste, s'est laissé endoctriner, parce que, avec son impétuosité native, elle s'est précipitée en aveugle dans le travail et l'abstinence, la classe capitaliste s'est trouvé condamnée à la paresse et à la jouissance forcée, à l'improductivité et à la surconsommation.

(Lafargue 1969 [1880]: 136)

Because the working class, with its simple good faith, has allowed itself to be thus indoctrinated, because with its native impetuosity it has blindly hurled itself into work and abstinence, the capitalist class has found itself condemned to laziness and forced enjoyment, to unproductiveness and overconsumption.

(Lafargue 1907: 34)

In Lafargue's thought, the working class, 'proletariat', and the capitalist class are clearly separated, even though it is difficult to identify exactly where the capitalist class starts and where the working class ends. The working class is in fact also called 'the laborers', and the 'productive class', which leads to a counter-definition of the capitalist class as, paradoxically perhaps, people who do not need to work, do not produce anything, and 'devote themselves to the overconsumption of the products turned out so riotously by the laborers' (Lafargue 1907: 34). Yet by brandishing the notions of abstinence and unproductivity, respectively linked to the working class and the capitalist class, Lafargue reveals the processes of compensation that are involved in conceptualizing both consumption and labour. What he also uncovers is an undeniable surplus, represented in the overconsumption but also in the very idea of leisure. In fact, surplus constitutes another way of looking at garbage and at waste. In order to understand waste, one has to question the role of surplus, and has to put time and energy in the equation.

Radiant City is based on an acute sense of the surplus: surplus consumption, surplus inhabitants, etc. For Le Corbusier, there is a sense of the overflowing city, whether it is material objects, people, dirt, or garbage. He is not the only thinker to question and renegotiate the idea of a surplus in the interwar period. Georges Bataille exposes a similar problematic in his *Accursed Share*, where he claims that something has to be done with this surplus, since it cannot indeed be eradicated. In his fascinating study on Bataille, Allan Stoekl identifies the modalities of this surplus in society. In his reading, there is a need for any social system to evacuate a surplus of energy, in order to constitute itself as coherent and complete. While Stoekl speaks mainly of energy in the sense of fossil fuels as opposed to solar energy, Bataille's energy could be considered as much more polymorphous: it could in fact be human energy, and its surplus would be what is left over after the hours of labour. This could be verified by Stoekl's interpretation: 'The inevitable limit of the system – economic, ecological, intellectual – always entails a surplus that precisely *defeats* any practical appropriation' (Stoekl 2007: 35). The key to Bataille therefore resides in the impossibility of practically appropriating the surplus: it could be argued that leisure, whether it takes the form of labour or of a complete idleness, fulfils that purpose.

If leisure is a given, then why not accept it as just another by-product of modern, industrialized, developed societies? Perhaps leisure and waste need to be *re*thought, thought anew, as intrinsically related and not necessarily reprehensible. What if the answer was not to waste not, but to waste better, to waste consciously?

In the meantime, it is undeniable that many successful endeavours on the subject of waste have a lot to do with leisure time; there are banquets with '7 × 70 The Meal'. Varda's documentary is a long leisurely walk on the subject of gleaning, and she does meet a few people who do sustain this lifestyle while working full-time jobs. Even then, we may need a communicative approach to garbage to spread awareness of the problems involved, and the partial solutions at hand. Artistic works such as Lucy and Jorge Orta's go a long way in familiarizing the public to environmental questions. An example of this is the exhibition 'Food/Water/Life', which took place at the Parc de la Villette in Paris in 2014, promising to reach thousands of children. There remain some vocabulary issues: I have said previously that the Minimum House is a comparable concept to that of *Zero Waste Home* or *No Impact Man*, but how can we explain that from the 1930s to the 2000s, people who care about the environment and the city have moved on from derogatory enough terms (minimum) to completely unpopular ones such as 'zero' or 'no'. Why does sustainability or green living have to be associated with extremes? It is unreachable, as both experiences prove, and even as a goal, it is not very attractive. After all, we do need a surplus, in the form of leisure, gardening, gaming. Zero cannot be marketed as something fun, it is a word they put on Coca-Cola when they want us to think it is better for us. We live in a world of low-fat and sugar-free. If the middle-class is what a post-Marxist society strives for, perhaps it is time for a middle-garbage, where neither dirt poor nor filthy rich would produce too much trash. If we are constantly being told by public figures, the media, and governments around the world that we are all part of the middle-class, it is revealing that, paradoxically, there seems to be no middle-class garbage, somewhere between the dirt poor, who consume trash, and the filthy rich, who produce too much of it.

Note

1 While Johnson does claim to work professionally full-time now, a great deal of her conception of a 'simple life' could be considered to be more time-consuming than what most people can afford to do on a daily basis: canning, shopping in bulk. In the previous quote, she admits to having spent a significant amount of time driving around the region to find more variety than in the bulk section of her local store, something she could hardly have done if she had not been a stay-at-home mother. Most of her lifestyle indeed seems manageable while working full-time, but the transition to zero waste is in fact the time-consuming part.

10 Under the spectacle

Viewing trash in the streets of Central, Hong Kong

Anneke Coppoolse

There is something about trash.[1] Like in any big city, in Hong Kong – specifically in areas that are both commercial districts and densely populated locales – it is produced at a constant pace. And while much of this city's discards first land in the streets, generally (perhaps paradoxically) – as space is limited – the government's attitude in dealing with it orders instant collection and concurrent 'hiding' in dumpsites. That is, '[t]o help keep Hong Kong tidy' (Food and Environmental Hygiene Department 2014), a large and visible workforce is sent out, daily, to deal with whatever people trash, from leftover take-home meals to complete furniture items. Subsequent to this apparent efficiency, just over half of the territory's total municipal solid waste is still landfilled (Environment Bureau 2013: 10). This corresponds with what Gay Hawkins (2007: 348) calls the imaginary of tidy cities, where the hiding of trash is vital to 'the maintenance of distinctly modern classifications and boundaries and distinctly modern ways of being'. Seemingly 'naturally', the government has equipped itself to cater for what David Boarder Giles (2014: 98) has labelled the 'bulimia of late capitalism'.

Capitalism – especially its late variant – has, like the imaginary of tidiness and order, a large 'visual suspense'. While some mention a shift 'from the economy of material production to an economy of signs and symbols' (Lash and Urry 1994, in Ma 2008: 64), and others understand visuality to have become 'one of the prime motivators of consumption' (Mirzoeff 2002; Mitchell 1994 in Ma 2008: 64–5), under late-twentieth century capitalism a preoccupation with 'surface appearances' is simultaneously said to inhibit 'a deeper understanding of underlying material forces', which can be seen to be 'symptomatic of false consciousness, alienation, and the workings of the market' (Zurier 2006: 7). Regardless, the buying and selling of things, as well as the things themselves, are spectacularized to great effect, in consumer society. And Hong Kong being one of the world's global cities, its flows of goods, money, and people – in addition to the modern urban landscape that lodges these flows – have contributed to the drafting of a spectacle of speed and flash (McDonogh and Wong 2005: xiv). The bulimic effects of Hong Kong's flows, however – even though these effects can be seen to force a spectacle of their own – have been largely left out of the picture. While what could be called a 'spectacle of trash' is never entirely ignored, as any big city is imagined to have some grunge – especially Hong Kong – overall, in the narratives and images of Hong Kong as global city, the flows of trash it nurtures are 'hidden', or at least are not given the attention the city's other flows receive.

The consequences of the rather rapid rendering invisible of trash in Hong Kong, are, together with the scale of it, an important issue. That is, if nothing changes, the landfills will reach full capacity by 2019 (Environment Bureau 2013: 7). And despite the proposal of the *Blueprint for Sustainable Use of Resources 2013–2022* (Environment Bureau 2013) – including various suggestions and policy and legislation changes for waste reduction at the source, and plans for the enhancement of waste-related infrastructures – trash will not simply disappear. As Gay Hawkins (2007: 350) states, '[a]ll cities, no matter how efficient their waste management, cannot hide the excesses of consumption'. And, in Hong Kong – its densely populated, urban and distinctively vertical side, that is (as Hong Kong is a territory that also contains remote islands, rural areas and country parks which I will not focus on in this chapter) – these excesses particularly reveal themselves at the level of the streets.

From a wider view of excesses and spectacle, on which I elaborate in the first part of this chapter, I move to the streets of one of Hong Kong's most iconic places of commerce and flow: Central. In places such as Central, order and hygiene are desired conditions of the modern urban landscape, advancing a 'smooth running of things' (Žižek 2006: 17, in Moore 2012: 781) towards a spectacle of flows. Yet, trash is regardless, unavoidably and at times uncomfortably present in such landscapes. As Hawkins (2006: vii) notes, trash exists not only at the 'ugly, shit end of capitalism', but also in 'paradise', exposing all 'yearning for purity as doomed to failure'. While I do not intend to argue for a spectacle of trash, in this chapter I look at how trash is present in a place of presumed tidiness, and what it can say about its context of global flows. I articulate in an ethnographic account of perspectives of the everyday in the streets – specifically those of one of the area's trash collectors, Shandong Lou[2] – a (visually inclined) story, however fragmented, of a different and underexposed side of Hong Kong's central business district.

Flows and spectacles, tidiness, and that which isn't tidy

Hong Kong, a global place since its emergence and cheered for its economic pursuit and distinct urbanism, is – along the same lines – regularly understood in relation to the idea of a 'world city' (McDonogh and Wong 2005: xiii) with 'nodal significance' (Siu and Ku 2008: 3) in a global economy. In this mindset of Hong Kong being a hub, it is often – even though it also embodies 'continuities of family, place, culture, action, and creativity' (McDonogh and Wong 2005: xv) – portrayed and analysed as a place or space of temporality, fragmentation, flow, and rootlessness (Sinn 2008; Kam 2010). Architectural historian Vittorio Magnago Lampugnani (Lampugnani *et al.* 1993: 11, in McDonogh and Wong 2005: xv) describes Hong Kong as 'an enormous monument to the transitory'.

And Helen Siu and Agnes Ku, in their edited book *Hong Kong Mobile* (2008), shed light on the various sides of Hong Kong as complex that 'may not be treated as a physically bounded crucible', but as a place that diverse populations relate to 'with economic interest, political commitments, social networks, and cultural imaginations that extend far beyond the limited physical space that is administratively defined' (Siu and Ku 2008: 10). Ackbar Abbas (1997: 4) – also staging Hong Kong as a global city, prefigured by its earlier colonial form – marks its position 'at the intersections of different times or speeds', particularly focusing on the cultural imaginations of the place. Although with great emphasis on its unique 'cultural' existence as a consequence of British colonial investment in, and Chinese postcolonial continuation of, its global direction, he famously calls for an understanding of the region's status of culture, as a 'culture of disappearance'. This 'disappearance' does not mean a vanishing of culture, but rather an appearance based on the 'imminence of its disappearance' (Abbas 1997: 7). That is, in light of the planned handover in 1997, Hong Kong – where transience and consumerism were accepted as main modes of being – took on 'a last-minute collective search for a more definite identity', that translated into various cultural forms (Abbas 1997: 4–5). It is, indeed, not just people on the move that add to Hong Kong's monumental transitory, other flows (of images, money, and goods), that contribute to its very particular (and particularly global) urban situation, and the images that it generates.

Further, as McDonogh and Wong (2005: xiv) explain, 'Hong Kong is no longer an exotic colonial outpost or a symbol of decadent capitalist modernity', as it had previously been imagined. It now evokes a place of 'speed and flash, known for its exciting films, its soaring buildings, and its frenzied pace' (McDonogh and Wong 2005: xiv). Abbas (1997: 2) explains this transformation in relation to a mutation in the capitalist system, drawing from Lash and Urry (1987), who describe a shift from 'organized capitalism' to 'disorganized capital', and Castells (1989), who details 'a movement towards the space of flows of the "information city"'. Indeed, Hong Kong's space of flows contributes to the construction of a specific (yet global) spectacle – understanding spectacle in the way Debord describes it, as 'a social relation among people, mediated by images' (Debord 2004: 7) that purpose the city and its life 'by the flow of commodities and their apparitions (advertising, cinema and so on)' (Highmore 2002: 61). This flow of commodities is seen to become ever more 'visual', while, as indicated before, late capitalism is moving from an economy of the production of things to an economy that is conditioned by visual signifiers (Lash and Urry 1994, in Ma 2008: 64) leading consumer societies (such as that of Hong Kong) to modes of consumption that are increasingly prompted by visuality (Mirzoeff 2002; Mitchell 1994, in Ma 2008: 64–5).

Taken that this context of speedy urban visuality and the spectacle of flows involves all sorts of movement of, and relations between, things, people, and images – and regardless of an increase in visual (and potentially non-material) consumption – the cycle of production and consumption, as mentioned before, also leaves (visual and material) traces other than 'flash'; traces that become flows of their own: flows of 'trash'. Indeed, in the gutter as well as at the centre of late capitalist and urban societies exists something that could be called a spectacle of trash (trash being utter rubbish as well as discarded commodities and everything in between). One of the most unwanted remnants of consumer society – in point of fact often regarded as 'matter out of place' (Douglas 2002 [1966]) – trash may have been deprived of prominence in the above-mentioned spectacle of speed and flash, but it is no less there. As Hawkins (2007: 348) notes with regards to her own city Sydney, 'despite sophisticated technologies for elimination, [...] waste is part of the landscape'. It is also widely accepted that not just the production and consumption of things (visual and otherwise) involves local, regional, and global flows of money, people, and goods. That what escapes its cycle, has become an economy in itself (Hawkins and Muecke 2003: x). Trash and filth, besides formally offering monthly salaries, because of the necessary management, collection, and processing, to more than 10,000 government employees in Hong Kong alone (Food and Environmental Hygiene Department 2012), can also be exchanged 'as recyclable resource, antique, tourist landscape' (Hawkins and Muecke 2003: x). Trash flows. And especially in Hong Kong, where urban density provides for its equivalent in trash, it often 'flows' to spectacular effects, in particular at local (municipal) and regional levels.

I refrain from laying out – in detail – the extent of Hong Kong's spectacle of trash, for that it only adds to the already existing understanding that Hong Kong's visuality is, in part, constructed by the vast flows of things, people, and images that it pertains to. Instead, I move towards understanding trash and its flows in relation to what Hawkins (2007: 348) describes as 'the modern urban imaginary [that] celebrated the tidy, the clean and regulated city that linked a public aesthetic to political order and civic consciousness'. This is an imaginary that is equally modern, yet seems near paradoxical to the above-mentioned spectacle of flows. The imaginary of tidiness may have contributed to the neglect of trash as vibrant flow of goods in the image of, in this case, Hong Kong as a speedy city. And so, as Hawkins (2007: 348) expounds as well, while modern ways of being involve a rendering invisible of society's trash, attempting to look at its presence in order to understand something more about a place is at least somewhat contradictory. But as John Scanlan (2013: 2) suggests, precisely the characteristics of modern life in which 'order, efficiency and perfection' are pursued reveal what he calls – borrowing the term from George Kubler (1962) – *aesthetic fatigue*, as there are places and events on the fringes – and, I would like to argue, also at the centre of the modern city (and of modern life) – where trash comes into view, and the kind of control envisaged by the modern imaginary dissolves. Or as Jane Bennett (2004: 348, in Hawkins 2007: 349) reasons, trash is 'materially recalcitrant'. It keeps returning (materially, 'sensuously', and, thus, visually), challenging governments' intentions to move it out of sight rapidly. It is of a constant 'nature', hence the flows persist.

The imaginary of a tidy city and the aesthetic fatigue it indicates, hint at what Dipesh Chakrabarty (1992: 541), in the context of colonial India's public spaces (bazaars and streets), explains as an unease between modern conceptions of hygiene and dirt and 'premodern' ways of living daily lives in the streets. In contrast to the modern market place – a market place that is exactly, in the case of Hong Kong, translated into the iconic image of business district Central and its 'soaring buildings' – Chakrabarty describes the bazaar and the streets as sites of 'economic activity', besides them being sites of 'recreation, social interaction, [and] transport' (Chakrabarty 1992: 543). Moving from Chakrabarty's context of colonial India's streets to the streets of today's Hong Kong – specifically those of its business district, Central – I suggest that exactly by looking at daily life interactions with regards to trash, interactions that are negotiated between the above-mentioned imaginary of tidiness and the system of consumption and production (including the visual persistence of trash this system breeds), an attempt can be made to see beyond (or underneath) the spectacle of the modern market place and related late capitalist imaginaries, to reconsider a view of (at least fragments of) Hong Kong's urbanity that is based on more direct and embodied (social) relations that are, in part, established visually.

Beautiful things

Here I shift to another mode of writing. My attempt at seeing beyond the spectacle, articulating a part of a story of Central's flows (and Hong Kong's urban life), through trash, draws on a larger visual ethnographic study involving trash and collectors in two different – yet iconic – commercial districts: Mong Kok in Kowloon and, indeed, Central on Hong Kong Island. For the purpose of this chapter, I take up one collector's experiences and views. I let myself be guided by Shandong Lou to 'see' Central (and surrounding areas) in a new way, not to overturn the spectacle as proposed in global narratives of flows, but to dialogue 'ways of looking' that are related to these narratives, to tell a different and, in part, visual story of a place largely known for its flash and frenzied pace. Shandong Lou is an experienced trash collector: a scavenger of sorts, as well as a 'sāu máaih lóu' (收買佬, literally translated 'collect and buy man'; a rag-and-bone man, historically known by this name in Hong Kong culture and that of the extended Guangdong Province). He recycles, apart from plastic, pretty much everything there is to recycle, and can be seen to work, therefore, as part of a (complex and layered) milieu of trash where discarded things are adopted back into the cycle of production and consumption. That I only follow him, and no one else, scopes this chapter and limits the extent of my focus. The chapter should therefore be understood as such: as elaborating only fragments of a larger 'trash-scape'. I leave out the various kinds of contracted collectors who can be understood to be (when on duty) directly working towards the imaginary of a tidy city, contributing to a flow of trash that often lands it straight into dumpsites.

I have known Shandong Lou for approximately nine months and after my interpreter[3] and I had befriended him, he soon became one of the contributors to my research. Consequently he started, at times, taking around a small camera[4] when working, recording over four hours of video, which we discussed afterwards, in two extensive interviews. In the following, I pick up, from his recordings and interviews (but also from other less official chats and interactions with him), some examples of what could be called 'visual events', from where I formulate different segments of a larger story. Mieke Bal (2003) coins the term 'visual event' in the context of visual studies. She suggests that visual studies should be about 'visual events' rather than 'visual objects'. Visual events take place 'in different locations and with different human and artifactual actants' and so – besides the subjects and objects that contribute to the events – also their 'situations' should be understood (Rose 2012: 544): the places and contexts within which they come about, and the practices (of looking and of other ways of doing things) that contribute to the shaping of these places and contexts. It is a new 'way of looking', explains Rose (2012), which I – in part – have taken up in my ethnography, combining the idea of the 'visual event' with a study of things, people, practices, and place, in the urban everyday.

Let me begin with a 'visual event' that introduces some of the sensibilities and classifications Shandong Lou negotiates. It also reveals fragments of a social situation of trash that can be articulated in relation to the modern market place which Chakrabarty (1992) suggests is the desired site for economic activity today. It starts with a quote from Shandong Lou (below translated from Cantonese into English). It is recorded by himself as he is in the process of filming one of the crossings in Central where he 'works' every morning between 5 and 7:30, before Central's regular office population heads out to go to work. In the early hours of the day this crossing transforms into the centre of a small 'tin gwong hui' (天光墟) – a dawn market – where a group of (as I will elaborate later) mostly elderly people, who have known each other for years, come to sell scraps and other things accumulated in all sorts of ways. 'Maybe I'll film your beautiful things', says Shandong Lou to one of the other dawn market sellers as he moves his camera towards some clothes laid out on a bit of raised pavement, zooming in on a second-hand teapot after that. Later, in an interview, he explains that it is a 'jí sā wùh' (紫沙壺), a Zisha clay teapot. He once sold a similar pot at an auction for HK$100 (approximately €10).

I ought to clarify that the things sold at dawn markets are not all 'trash'. They are things that are, as Leung Chi-yuen (2008: 133) explains, bought and sold via translocal and even transnational systems, as well as between different tiers of informal economies in different areas of Hong Kong. The things are acquired (as has been explained to me by some collectors) in various ways, such as from homes that are being refurbished; from people who give their old things to sellers whom they happen to know; indeed, bought at other informal places of commerce; but also simply picked up from between the garbage. As the latter may border illegal practices, I never directly indicate whether an object (apart from paper waste, which is acceptable to pick up) has indeed been collected from the streets. However, I would add that it is commonly known that many of the things sold at dawn markets are acquired in this way. Furthermore, in Central – and more specifically just on the edges where more residential buildings can be found – it so happens that because of the large expat community, people move in and out of apartments at a high rate, discarding also, in high numbers, things that would still be regarded valuable by those selling at dawn markets. Indeed, the kind of trash I am referring to here and in the rest of this chapter, is that which Hawkins and Muecke (2003: x) allude to as trash that can be exchanged 'as recyclable resource' or even 'antique'.

Shandong Lou, fixing his frame for a few seconds on the teapot, does not stop at that. Having just commented, 'Maybe I'll film your beautiful things', he continues along other improvised stalls (see Figures 10.1 and 10.2). When, in the

Figure 10.1 Dawn market stall. Screenshot of video by Shandong Lou (Central, Hong Kong, April 2014).

Figure 10.2 Dawn market stall – display of shoes. Screenshot of video by Shandong Lou (Central, Hong Kong, April 2014).

interview, we ask what made him make these recordings, he says 'What? This is where we sell our things in the morning! This old lady's stall is beautiful. It is interesting to see how she has positioned the clothes and shoes.' The second part of his remark refers to the stall presented in Figure 10.2. Indeed, in both images, the ways in which the things are displayed – neatly folded and arranged – present an order of a kind that may not fit the order imagined for the tidy and regulated city, for that everything discarded is to be immediately moved out of sight in the name of public order. Yet, this exhibition of things – things that in other settings and situations (for instance, randomly lying around in the alley, next to some rubbish bags) would be taken as trash – has consequently helped them to over-come their 'trash status'. By their careful display, previously discarded things are adopted back into the cycle of consumption. The dawn market sellers reflect commercial practices in the off-hours of the day, literally in front of the more upscale shops and in the shadows of large corporations, presenting – visually – 'trash' in new ways.

Shandong Lou, calling the objects as well as the way in which they are pre-sented 'beautiful', adds another layer of articulation to their revisited visuality. He surmounts, in part visually, what Hawkins (2007: 348) calls 'distinctly modern classifications and boundaries and distinctly modern ways of being' – classifications based on binaries such as trash/not trash, unsuited to the public aesthetic/beautiful, but also valuable/valueless – concurrently moving towards a logic of consumption that, as previously explained (Lash and Urry 1994; Ma 2008), is largely motivated visually. The sensibilities and classifications Hawkins locates, are those that induce an 'exclusion' of trash, both through organized waste management systems as well as more embodied practices and aversions of individuals in modern society (Hawkins 2007: 348–9). Those Shandong Lou works towards and against, are – instead – based on the potential inclusion of trash by means of an appropriation of other modern logics; logics of display and consumption.

Before elaborating on the next 'visual events', I briefly return to the dawn market's relation to the established global economy that is serviced in Central's soaring buildings. Dawn markets for the most part suffer under Hong Kong's 'free market' policy (Leung 2008: 94); a policy that has contributed to the glo-balizing of the city. As Leung (2008: 94) notes in a related and highly detailed study of illegal street hawkers elsewhere in Hong Kong, even though such selling practices have a long history in the Guangdong Region, today they form an evolving response or even resistance to the growing formal economy. The hawker businesses that he describes are subject to 'active intervention of neolib-eral governance' (Leung 2008: 97) as they are monitored and constrained by various policies. After 1997 (the handover), the governing direction set out by the British was furthered, meaning that Hong Kong's development into a global city had induced even more strict regulations on, for instance, street hawking and selling (Leung 2008: 105). Such regulations echo Chakrabarty's (1992: 541) claim about the unease that was felt in colonial Indian cities towards the bustle in public spaces, where trash and economic activity were perceived to coexist.

Yet, that illegal street hawking and the buying and selling of things at dawn markets – which Leung (2008: 133) classifies as the lowest tier in Hong Kong's informal economy (with legal street hawking just one tier up) – is still happening, suggests a counter-presence of a sub-economy of trash and other things, in the case of the dawn market in Central, right at the heart of a place that is well regarded for its capitalist undertakings. It is a counter-presence that hints at an urban situation that is economically more diverse than Central's spectacle proposes.

Taking this argument further, towards the market's social position, Shandong Lou and the others form the human force of what Giles (2011) calls a 'social afterlife of things'. In return, this afterlife contributes not just to a sub-economy of trash, but also to the formation of a distinct group of people: people who, during the day, scatter over the area (and beyond) – most of them (if any) in low-wage jobs or, indeed, collecting cardboard. It is also a group that is, however 'locally transient' for the nature of their work, quite continuous. Shandong Lou, in the interviews, has indicated various times, when others at the market are in view, that he has known them for decades. It seems important to elucidate that he, as his name implies, is a migrant from Shandong Province in Mainland China. He came to Hong Kong in the late 1970s and has since been living and working mostly in Central and neighbouring Sheung Wan. He is, although a migrant, not transient as understood in the more spectacular sense of Hong Kong. Yet, he still operates 'at the intersections of different times or speeds' (Abbas 1997: 4) in response to global flows, not just of things, but also of people. Indeed, while his fellow market sellers are, like him, quite connected to the place in which they work – fixed – a notable part of his clientele is transient and transnational. That is, a relatively large number of people buying at the dawn market are domestic helpers. The things they buy – often clothes and shoes – are, as Shandong Lou explains, regularly sent to their home countries, sometimes even for further commercial purpose. The dawn market can, therefore, be seen as a small node in a sub-economic, translocal, and transnational flow of things, responding to, as well as resisting, its dominant counterpart. And, while it brings together people picking, selling, and buying, when taking up on some of its visual events – understanding them in relation to today's favoured tidiness and modern logics of consumption – it suggests a visuality different from those somewhat distant spectacles and imaginaries; it suggests an embodied visuality that involves things and people, and the place within which it occurs.

'En route': cardboard and other things

Shandong Lou is 'one of a kind'. He collects – different from most other people who earn money by selling collected trash – a wide variety of things. The things he sells at the dawn market are usually 'daily necessities': clothes, kitchen utensils, small furniture items, and the like. But throughout the day he also gathers things that are not options for second-hand trade at the market: cardboard and other recyclables, and 'antiques'. Recyclables can be sold at recycling shops, and antiques – if he can get a hold of them – he may send to auction or try to sell to antiques shops. At the same time – being familiar with his area and with the people in it – he regularly receives things from people directly, as well as passing them on in this way. For electronics he 'knows a guy', for instance. While most people – often elderly ladies – keep to just cardboard, I have come across a few others collecting in ways similar to Shandong Lou: people who collect close to everything. One of them once explained that it depends on people's knowhow and their physical state. The physically weaker go around collecting cardboard. Or they bind up white Styrofoam crates that are often used to transport goods sold in wet markets – crates that are picked up in the night, by delivery trucks returning to Mainland China. Those people who are a bit stronger, however, and have, for instance, a background in construction or refurbishing, can afford to pick up bulky items and dismantle them accordingly (Personal interview with a collector, Mong Kok, 2014).

Returning to Shandong Lou, who, while on the move, allows me a 'peek over his shoulder' at what he finds, what he sees, and where he goes while in search of recyclables. Below, I elucidate two 'events' that visually exemplify and concurrently explain some of his movements and chance encounters with certain things. In these examples, he can again be seen to respond to certain existing logics of production and consumption, and to resist others. Shandong Lou works between two areas – Central, which is the more cosmopolitan part of Hong Kong's commercial district, and neighbouring Sheung Wan, which is quieter and known for its antiques shops, boutique hotels, and upscale restaurants as well as residential quarters. Sheung Wan also has a few recycling shops because of which the sidewalks between this area and Central see much traffic of recycled materials on trolleys. On one afternoon, Shandong Lou walks from higher up the hill, in Sheung Wan at the border with Central, to the park opposite the Man Mo Temple. Different from other moments in which he finds cardboard – which he then piles up or binds together – he comes across a cardboard box with a discarded Chinese ceramics book in it, stained. All of this he video-records. Being close to the park, he initially takes the book not just as paper waste. He moves to the park to glance over the pages. But, more importantly, he wants us to see inside the book too – he video-records every single page (see Figure 10.3).

Coincidentally, as he is making these recordings, we call him to pick up the camera. He tells us to come to the park and as we arrive he says that he found the book behind one of the antiques shops, across the street. Indeed, as I explained above, there are numerous antiques shops in Sheung Wan and Shandong Lou, having lived in the area for many years, has concurrently developed an interest in antiques. In the past, Shandong Lou makes clear, he even sold antiques in shops and hawker stalls in the same area. To what extent he sold antiques, and what exactly happened between then and now, remains unclear. However, while he tells us that throughout his life he has had various different jobs, from factory worker in the early days to security guard later on, the selling of things in the streets seems a more constant occupation. Although for this chapter it is not quite necessary to dig much further into his personal life, I have

Figure 10.3 Discarded Chinese ceramics book. Screenshot of video by Shandong Lou
(Sheung Wan, Hong Kong, March 2014).

found that – having spoken with many other collectors over a period of a year
and a half; people who are usually considered part of Hong Kong's lower classes
– a relative number of them share similar stories: of migration (from the Mainland, but also from elsewhere in Asia), often followed by comparable economic
struggles that can be understood in relation to Hong Kong's transformation into
a post-industrial global city. Indeed, post-industrialization in Hong Kong
affected particularly low-wage workers (Lee and Wong 2004: 262–8) – a topic
that I will not further expand in this chapter, as it is highly complex and requires
detailed elaborations. Regardless, although Shandong Lou's recordings of the
stained ceramics book are by no means direct indications of Hong Kong's recent
history and its mutation from a city of imagined 'decadent capitalist modernity'
into a place of 'frenzied pace' (McDonogh and Wong 2005: xiv), they do offer
entry into talking about these matters from the level of the streets.

Considering the above-described 'visual event' in relation to the need to
render trash invisible in the contemporary city, and coming back to what
Hawkins (2007: 348) defines as clear modern boundaries between what does and
does not fit the modern mode of being, the moment in which Shandong Lou
finds the stained book and starts looking through it also helps to reconsider the
relation between antiques sold in upscale shops, and the trash that is dumped at
their backdoors. The antiques shop operates based on modern classifications of
'commodity/rubbish' and 'to be presented/to be discarded', displaying and
'hiding' things accordingly (filtering 'valuable' items via the shop window
through the front door, and 'valueless' items through the door in the back).
Shandong Lou picks up that which has been discarded and revisits its category,

visually. He values it for what it represents – imagery of antiques, in part based on his personal experiences of living and working in a place of certain spectacle. Even though, that day, he confirms that he will not try to sell the book as a book – as it is, indeed, only sellable as paper waste (accepting the predefined modern category of paper waste) – his way of looking at that which is featured in it highlights particular aspects of both him (being 'one of a kind') – his interests – and the environment in which he operates.

Shandong Lou's interest in antiques and his surroundings also comes to the fore in many other recordings. He captures antiques in shops and behind shop windows, and takes the camera around Cat Street: an antique street he mentions a lot, because he worked there in the past (see Figure 10.4). Even though antiques have his interest, and while he also sometimes sells them himself, on various occasions Shandong Lou has indicated that to him all things he collects – from antiques to paper waste – are the same: 'they all have the same destination'. Shandong Lou – currently selling without a shop – does not filter the things he sells in the way the antiques shop does: antiques through the front door, trash via the back. Shandong Lou, as he revalorizes discarded things, does not adapt to modern binaries such as 'to display'/'to hide'. Instead, he keeps to his sub-economic network of things and people and distributes things accordingly. He becomes his own shop – a true 'sāu máaih lóu' (收買佬). And this is perhaps best illustrated by the many jade necklaces that he wears around his neck at all times, on display – in case anyone is interested in buying one.

Shandong Lou's revalorization of particular objects also implies something more subjective. As Hawkins and Muecke (2003: xiii–xiv) explain, expelling and discarding 'is fundamental to the ordering of the self'. And, in the same vein, a revisiting of expelled and discarded objects can be understood to have

Figure 10.4 Objects on display in Cat Street. Screenshot of video by Shandong Lou (Sheung Wan, Hong Kong, March 2014).

such effects. That is, the management of trash is, to Hawkins and Muecke (2003: xiii), 'deeply implicated in the practice of subjectivity'. Practices of collecting and organizing waste (in all their different forms) are involved with the ways in which individuals cultivate particular 'sensual relations with the world' (Hawkins and Muecke 2003: xiv). Shandong Lou, reconsidering distinctly modern boundaries between trash and order, can be seen to have – undoubtedly, yet to an extent by necessity – revisited relations between himself and the world around him too. Being 'one of a kind', he has diverted from a certain expected practice of subjectivity in the spectacular as well as the tidy city, which leads me to elaborate on another 'visual event'. That is, while Shandong Lou calls on his sub-economic network of discarded matter when dealing with all he has collected, certain relations between his senses and his environment cannot be ignored. And so, the next event manifests how Shandong Lou and his things are 'seen' and responded to, sensorially, by certain others.

Dealing with the bulimia of late capitalism in his own way, Shandong Lou, understandably, stores – for longer or shorter periods of time – the things he collects in various places. He keeps his things often in the alleys, out of sight. However, he will also leave them on the pavements (always making sure pedestrians can still pass). I should add that, in Hong Kong, pavements are regularly used as places for the temporary storage of things – piles of parcels that are to be delivered; cardboard-loaded trolleys that are to be brought to recycling shops, etc. One of the locations where Shandong Lou sometimes chooses to park his things is a specific corner in front of a jewellery shop. One morning, after the dawn market has wrapped up, we are chatting to Shandong Lou at this corner. Shandong Lou, working and talking at the same time, is arranging his collected cardboard before taking it to the recycling shop, while his other things are temporarily parked – rather neatly – at the corner of the street, on the road-side of the pavement. After some time, one of the shop assistants comes to the door and tells Shandong Lou that her boss is on his way, so he should move his things. Shandong Lou explains that the boss does not like to have his things in front of his shop, because he thinks that it keeps customers from walking in. The visuality and materiality of the collected materials Shandong Lou is working towards distributing, disturb – in the mind of the boss – 'the smooth running of things' (Žižek 2006: 17, in Moore 2012: 781) in his shop. It may influence potential customers' sensory response to the place – Shandong Lou's 'pile of junk' may visually and materially obstruct, but also sensorily get in the way of, his business.

This event exemplifies exactly the modern need to render trash invisible. It is untidy, disorderly, and – as per the boss's rationale – opposes the kind of visuality of things that is necessary for shop owners to keep business running smoothly. It is, thus, not just Shandong Lou and with him other trash collectors and dawn market sellers, who negotiate the visuality of trash by its material appearance at the level of the streets. In the name of order and hygiene (subsumed under a system of production and consumption that is to a notable extent motivated by visuality, while its exact working is consequently confused by a preoccupation with 'surface appearance'), those who (think that they) are not directly engaged with trash – those who are confronted by it 'unwantedly' – respond to it in equally sensuous ways, informed by a system of norms and 'values': a capitalist system that suggests consumption but not its glut.

I take the above-mentioned example of the unwanted views of, however neatly packaged, 'trash' as my lead to a brief conclusion. Indeed, trash is often seen as something that disrupts socio-spatial norms (Moore 2012); norms that are – as argued across this chapter – predefined by certain modern ideas of hygiene and order engrained in the individual as well as in the larger body of society. In the tidy city and through a spectacle of flows, trash is largely ignored by means of a certain imagery that mediates social relations between people and things (leaving out those relations to things that are unwanted) or it is attempted to literally hide those things that are considered 'out of order'. This all happens to more extreme effects in areas such as Central, where tidiness and order are tied to global recognition and economic prosperity. At the level of the streets, however, 'ignoring' and 'hiding' is to be executed at a manual pace and trash can concurrently – in that period between disposal and collection – also be picked up by collectors such as Shandong Lou, who see in it still some value and who concurrently find opportunity to 'display' it differently.

Matters of trash in everyday encounters, the way they are negotiated, and the new kinds of (visual) relations these negotiations propose – while they, at times, are seen to obscure the smooth running of business (as in the example of the jewellery shop) – do not directly disrupt the larger flows of things. Instead, they generate moments in their socio-spatial reality, that consent to 'interferences' with modern desired tidiness and commercial spectacle. In Central, while still subject to the global flows of things, people, and money that the place bears, trash (and its visuality) in the everyday proposes perspectives of a context that extends and details that of flash and speed: a context of more local and regional movement of things; a context in which people of certain (relative) fixity – in part – depend on and cater for those leading more transient and transnational lives (moving expats and domestic workers, in the case of Shandong Lou's involvement with trash); a context, thus, of an economically, socially, and also ethnically diverse urban situation. Trash also steers one to question certain accepted modern logics of production and consumption – especially when taking note of its visuality. The flows of things and people that global cities accommodate, can be dissected and revisited (and their image partly resisted) from the location of trash, and through the everyday visual events trash takes part in. By understanding the visuality of trash in urban Hong Kong not in terms of its spectacle, but in terms of visual events, fragments of another, more detailed, story of the global city can be highlighted; a story that, while maintaining its backdrop of spectacular flows, is compiled of certain embodied, sensory, and visual experiences of urban life; a story that is not informed by an assessment of visual objects alone, but that is supplemented by the recognition of that what happens between objects, subjects, and locations.

Notes

1 I use 'trash' rather than, for instance, 'waste' mainly for semantic reasons. 'Waste' connotes an instant loss – something that has 'gone to waste' or that has been 'wasted' – while much of the 'stuff' that is 'trashed', and that I deal with, is (or could have been) revalorized in different ways.
2 Shandong Lou is one of the participants in my larger research project and has requested to be referred to as 'Shandong Lou'.
3 Mastering just about some conversational Cantonese, I usually hit the streets together with an interpreter. While this has obviously added another layer of mediation to my research, I have found the process of translation and the related conversations that emerged between us, at the same time productive and informative.
4 I provided Shandong Lou with a small mobile phone-sized camera that could be positioned in a case in the middle of his chest. He could choose to take it out or leave it in while recording. We did not give him much instruction as to what I wanted him to record. We simply explained to him that I am interested in anything that he finds important to record in relation to his work with 'trash' (and other things) and the way he sees and navigates the urban space in which he lives and works.

Part III
Abandonment

11 Geospatial detritus

Mapping urban abandonment

Joshua Synenko

Introduction: defining abandonment

Urban abandonment has long been a favourite subject in discussions about the aesthetics of waste. The haunting detritus of urban industrialism at Gunkanjima Island (Figure 11.1), for example, is just one site in which this aesthetic has been prominently expressed in recent years. However, this particular chapter does not investigate the obsession with ruins that is widespread, but instead asks more specific questions about how this obsession is communicated or transmitted. The main premise of this work is that ruin photography as a genre of art has been forcefully supplemented by popular formats of remote entertainment like Google Earth.

Such imaging technologies enjoy modes of access to abandoned space that invite us to reconsider Marc Augé's demand for an 'anthropology of the near'

Figure 11.1 Gunkanjima Island (2010) (courtesy of Jordy Meow).

(2009: 7), as it was through this demand that Augé sought to establish a new spatial category of transfer zones – airports, railways, and exchange hubs. These 'non-places', as he called them, tend to be designed without any substantial ties to broader histories and identities; they function instead by separating lived environments from each other as points on a map. Sites of urban abandonment share this description because of the reliance upon remote forms of witnessing that digital interfaces support. Above all, non-places delineate highly individualized experiences of place, and these experiences spotlight not only the risk, but also the potential for creating meanings that irretrievably lose the collective memory of the site under consideration. By extending Augé's concept, the abandoned site becomes implicated in a process that is marked by 'an absence of the place from itself' (Augé 2009: 85). I therefore suggest that Augé's definitional exercise is still quite pertinent, as it refers to the 'fleeting images' of a computational medium that 'enables the observer to hypothesize the existence of a past and glimpse the possibility of a future' (2009: 97).

The individual therefore mounts a challenge to the unifying memory of the collective, and establishes what Thomas Elsaesser has described as *île de mémoire* (2009). That is, by pushing the melancholic experience and its potential for unifying collective responses toward 'a certain extraterritoriality' (2009: 33), Elsaesser proposed a ritual of remembrance where the individual as such becomes an agent of its own experience; where, in other words, individuals create responses in which to supplement the forgotten memory of the abandoned spaces that they encounter remotely. Beyond the obdurate motifs of so-called haunting experiences, which tend to result in a mere collective gasp of fascination, I argue that this new designation of urban abandonment establishes relations with individuals who are able to bear the mark of its alterity. To put it another way, this new designation makes individuals capable of responding to the conditions of abandonment in ways that further acknowledge the specificity of those spaces, whereas collective responses facilitate little more than specular enjoyment. With the digital preservation of Gunkanjima Island, for example, we have created an island of memory for the individual.

Before investigating the shifts in perspective that digital formats of memorialization require, it is important to acknowledge that urban abandonment happens for many reasons. For example, a recent study (Tonkiss 2013) describes such abandonment in relation to the practices of austerity urbanism, which is composed of last-ditch efforts to force a neo-liberal agenda upon the world's most depleted cities. The Detroit water crisis that began in 2014 is a recent example of the impact that such divestment can have for a vulnerable population. On the other hand, as solidarity initiatives respond to this particular crisis against the perception that abandonment is the only course of action, the romantic appeal of Detroit in its ruined state continues to be advanced by those calling for its preservation. Such attitudes have situated Detroit at the forefront of a new cultural memory of divestment, in a situation where the city itself risks becoming a spectacle of its exalted past. Contributing to this romantic image of urban ruin is the recent publication of picture books in which Detroit is conspicuously featured (Austin 2012; Austin and Doerr 2010; Marchand and Meffre 2010; Levine and Moore 2010). Broadly speaking, this genre of photography is devoted to the representation of irreparable built environments. Historians, for their part, may find inspiration from these expressive formats to reinvestigate Detroit's role in America's industrialization, and perhaps even to develop a cautionary tale of the larger economic circumstances that led to this particular outcome. The visual material that often defines this genre, however, tends in most cases to decontextualize the causes of ruination, and it therefore shies away from constructive forms of social engagement. A notable example in this regard is Charlie LeDuff's *Detroit: An American Autopsy* (2013). The basic premise of this work is that America's abandoned ruins foreclose hidden truths of post-industrial life that can be revealed in turn through a narrative of personal discovery. LeDuff's work in particular is intended to excite and flatter the reader by inviting them to follow the dangerous and heroic story of his photographic adventures. As a further extension of austerity urbanism, therefore, the genre of ruin photography participates in the musealization of cities for the benefit of consumers.

Global urban abandonment has become an important field of investigation for the memory culture industry. Rebranded in recent years under the label of *dark tourism*, this particular industry works to assuage the casual interest of the viewing public in sites defined by death, entropy, ruin, or haunting (Bennett and Seaton 1996; Foley and Lennon 2000; Dann and Seaton 2002; Sharpley and Stone 2009, 2012; White and Frew 2013). By spotlighting frontier-like spaces that get excluded from the itineraries of the tourist industry, this new market initiative came about in an effort to sustain capital investments in the enterprise of a specifically *global* collective memory. Dark tourism, in any case, appeals to a small subset of consumers that is unsurprisingly located in 'countries and regions enjoying the highest rates of economic growth' (Sharpley and Stone 2012: 2). But this pooling of consumers raises a host of additional questions: from the ethics of repurposing these sites to the practicalities of gaining access to sites. To briefly illustrate just some of the issues that might be raised, I examine a proposal by ZA Architects to build a massive infrastructure at the site of the 1986 nuclear disaster in Chernobyl (ZA Architects 2011). From the outset, the promotional material for this project claims that it will address 'the socialization of the territory, development of infrastructure elements that facilitate tourism and scientific activity, [the] development of industry [and] environmental protection' (ZA Architects 2011). The proposal goes on to state that 'the paramount attention is devoted to tourist infrastructure' (ZA Architects 2011), though it remains unclear in their description whether the project as a whole is defined by political, educational, or commercial interests. On the other hand, a further set of questions has been raised (Bennett and Seaton 1996) regarding the impact such tourism makes in reflecting consumer *behaviours* – including the consumers who end up visiting the sites as well as those who are merely fascinated by the images. As an 'experience economy' (Pine and Gilmore 1999), the curatorial innovations that mould these sites to the needs of its audiences appear to emphasize the transmission of emotional or affective material. In other words, by 'privileging the 'visual' and the 'experiential' over historical rigour' (Sharpley and Stone 2009: 13), developers cling only to the most popular formats of representation, complete with alternative curatorial methods and new media applications. In the process, they dissolve the informing logic of the traditional museums with an exploration into geographic objects for the purposes of entertainment. As Sharpley and Stone (2009) argue, the trend within dark tourism is to recreate built environments as *memento mori*, in which affectively charged forewarnings are provided regarding the apocalyptic certainty at the heart of modern civilization. For Andreas Huyssen (1994), however, the affective dimension of these productions is less a symptom of remembrance (i.e. as forewarning), as it is a general indication of collective forgetting. Huyssen writes, 'the museal gaze expands the very shrinking space of the (real) present in a culture of amnesia' (1994: 35).

If the true motivation of cultural memory proves to be an impossible demand for knowledge, Huyssen can offer a viable alternative to Sharpley and Stone's claim that abandoned space tends to leave the impression of a general forewarning. Huyssen's perspective emphasizes the underlying motive for returning to spaces of abandonment, noting that the distinguishing feature of these spaces is often contravened in the very act of remembering them. As such, this particular framework is helpful in a further sense when it comes to determining the viability of retrofitting abandoned spaces with infrastructures to facilitate memory tourism. In some cases, it may be more beneficial to leave abandoned space alone, particularly when it comes to sites and geographies that are associated with unresolved conflict. In fact, significant benefits may be associated with resisting commemoration, if not the very desire for resolution itself. The belated German history of the counter-monument is a notable example of this struggle, as Germans refused until the 1970s to witness the excavation of sites connected with the genocidal crimes of the Nazi era (Synenko 2014; Young 1994, 2002). Designers drew inspiration from the revelation of the *Gestapo Gelände* in Berlin to create a new paradigm of memorial aesthetics, in which the negative space of the archaeological site became a way to acknowledge the crisis of identification that results from remembering crimes that are perpetrated in one's own name (Young 1994: 81–91). However, if we compare this German culture of memory with other geographies that have been implicated in genocidal violence, even within Europe, we find that the sites of atrocity in question may not provide such concrete remainders of the past. During the Yugoslav wars, for example, a genre of political violence was invented (Berman 1996) to analyse the targeting of built environments for total destruction. Known as *urbicide*, this analysis describes a form of violence in which cities themselves become the subject of eliminationist policies during campaigns for ethnic cleansing. In these examples, the perpetrators destroy not only the buildings that house civilians but also their critical infrastructures, like water and electricity, as well as their cultural symbols, such as monuments, which signify particular relationships to the areas in question. The destruction of the Bosnian *Stari Most* is widely considered to be the first subject of study into urbicide (Berman 1996; Coward 2008). Defined by Coward as a 'symbol' of ethnic cleansing (2008: 17), the practice of urbicide is not merely a by-product of genocidal violence but is rather its own category. Though 'a criminal deviation from the norm' (Coward 2008: 23), urbicide exceeds the provision of collateral damage and the rules of war in general. As an expanding field of study, the urbicide model has been applied more recently to conflicts in Chechnya (Coward 2008) and to Palestine (Abujidi 2014; Graham 2003). The impact of this work for the question of memory culture is significant, because it addresses whether legitimate and effective strategies of memorialization are even possible in regions affected by this level of violence. If total destruction is all but assured, by what operation can such violence be remembered?

One of the more popular ways that atrocities tend to be remembered is through public awareness campaigns, many of which use geospatial technologies like Google Earth to secure evidentiary demands. However, critical research (Parks 2009; Kurgan 2013) has suggested that many of these campaigns work in tandem with geopolitical agendas, and in some cases they are mobilized with the aim of distracting the public from larger concerns. Laura Kurgan (2013) has broadly historicized the public availability of satellite images from the 1990s onward as being instrumental for providing justifications of Western military intervention. Whether in Bosnia or in Grozny, the satellite images were made available during these conflicts mainly by reports in the news media. They were represented as highly authoritative documents with irrefutable proof of ethnic cleansing. Yet Kurgan indicates her scepticism towards these claims. The maps in question, she writes, 'come to us as already interpreted images, and in a way that obscures the data that has built them' (Kurgan 2013: 26). Guided interpretations of specialized visual material might be excused for the greater good of bringing genocidal crimes to public attention. Kurgan, however, describes how these particular initiatives can result in an entirely new level of injustice. The prime example of this injustice according to Kurgan is Colin Powell's use of satellite images and aerial photography in his 2003 presentation of evidence that Saddam Hussein hid chemical weapons from UN inspectors – a claim that was later proven to be false. Lisa Parks (2009) describes a similar strategy of soliciting a global public response to the unfolding tragedy in Darfur in 2005. She cites a joint effort by Google Earth, newly minted at the time, and the Washington Holocaust Museum to create a 'Genocide Prevention Mapping Initiative' (USHMM 2007). Intended to represent acts of genocide in real time, this effort has been described in promotional media as a 'living memorial' (USHMM 2007), one that utilized the moral authority of the Holocaust Museum in conjunction with Google Earth's technology, to confront 'threats of genocide and related crimes against humanity' (USHMM 2007). Noble though it may have been, Parks describes this initiative as an elaborate publicity stunt which was aimed at manipulating the emotional response of a global audience. The implication is that such efforts invariably result in perpetuating specific narrative and genre conventions that thoughtlessly accommodate prejudicial motifs to fabricate a sense of shock and moral outrage. The project's lack of success in pressuring Western governments to action was only compounded by its lack of a critical historical perspective in which to understand the geopolitics of the region, to provide insights into the conditions for the emergence of the conflict, or even to analyse facts on the ground (Parks 2009). As such, though the geospatial technology in question brought public attention to the events in Darfur, it retained a scopic gaze that steadfastly reflects its ever-present military function.

Undoubtedly, the genealogy sketched above pivots on the surging public availability of satellite images following Google Earth's inauguration in 2005. The result of this initiative has been a ubiquitous cartographic interface that in many ways is utterly unique. For Kurgan (2013), the practice of mapping in the digital domain situates a new vantage point of unfathomable depth from high above, a position that devours and empties spaces that are otherwise defined by human activity and movement. Described by some commentators (Scharl and Tochtermann 2007) as the *geospatial web*, this particular medium acts like an abandonment machine, for it privileges a technological gaze that has the capacity of suspending movement across vast distances with considerable precision. If our hope is to understand the dynamic operation that this act conceals, we must go beyond simply applying the visual medium to demands for *evidence* of abandonment. In other words, the previously explored investigations into austerity urbanism, ethnic cleansing, and urban destruction should now be set aside for a new mode of inquiry. Because the interface produces distortions of scale and perception, Kurgan, for one, insists that such investigations be replaced with an impact assessment of relinquishing the most entrenched category of the human subject: our 'sense of stable and fixed location' (2013: 15). For Kurgan, in other words, the geospatial web puts 'the project of orientation – visibility, location, use, action, and exploration – into question' (2013: 17). It therefore challenges our ability to enforce a reliable difference 'between the image as a "site of activity" and a "memorial"' (Kurgan 2013: 30). Investigating the memorial function of the geospatial image will thus require a different approach to assessing the latter's cultural value – a determination that could challenge or even replace the standing cosmopolitan vision that is representative of work by Denis Cosgrove (2001, 2006), for example. Now, Cosgrove rightly insists that the millennial history of maps and globes contain relevant indications of the present-day value that these objects enjoy. For instance, the images that were produced during the lunar mission in 1972 cemented a long-standing hope or belief in the 'unity and harmony' (Cosgrove 2006: 13) of all peoples who inhabit the Earth. This belief found its limit only in the perennial conflict between innovation and environmental destruction, that 'delicate interplay of forces and processes operating within the material world' (Cosgrove 2006: 14). Today, however, Cosgrove's assertions must be reconsidered beyond this ideological divide. Kurgan (2013) states, for example, that today's global imagination has been repackaged and reconstituted from raw material provided by a collection of satellites, material stitched together into a seamless image to create a *semblance* of unity. While 'the maps presented here record situations of intense conflict and struggle', echoing Cosgrove's assertions, they also signify 'fundamental transformations in our ways of seeing and of experiencing space' (Kurgan 2013: 14). In fact, the dissonant feeling that Kurgan describes has prevailed, in many ways, over any lasting sense of unity or harmony.

This chapter examines the transformations mentioned here by drawing from two unique examples of urban abandonment: from the surreptitious mapping of Michael Heizer's *City*, a sculpture hidden deep in the Nevada desert, and that of Gunkanjima Island in Japan, which has recently become a tourist destination for the virtual traveller. I illustrate the impact of the new spatial media on our shared sense of comprehension and attunement, noting that in both examples the desire to remember these spaces refers not only to their pasts or their conditions of abandonment, but also to contemporaneity. Indeed, the act of memorialization refers in both examples to a 'fantasy of proximity' (Kurgan 2013: 30) that tends to be at play in most varieties of remote witnessing.

Vertical space and the logics of abandonment in Michael Heizer's *City*

Convinced by the need to protect the originality of his ideas, Michael Heizer left the New York art scene in the 1970s to continue his production of earthworks in the solitude of the Nevada desert. He settled on a patch of land some 300 kilometres north of Las Vegas and east of Yucca Mountain, but long before the Mountain became a nuclear waste repository and the subject of ongoing contestation, including by Heizer himself (Knight 2009). Finding solace from the manufactured consent that he sensed was prevalent among his New York colleagues, which included Robert Smithson, Walter de Maria, and several others, Heizer began in earnest to build a structure so great that it would eventually challenge the art movement to which his work has long been associated. Known

Figure 11.2 Google Earth image of Heizer's *City*, south end (Map Data © 2014 Google).

today as *City*, this project is often regarded as the largest of any previous earth-works structure built before. It depicts an urban ruin composed from massive concrete slabs that have been situated in various 'complexes', distributed with geometrical precision along fabricated valleys and hills across roughly two kilo-metres of desert land. Precedents for this work can be found in Heizer's *Double Negative* (1969), a massive desert trench that he made entirely of empty or neg-ative space, and *Effigy Tumuli* (1985), which represents an attempt to repurpose the space of an abandoned mining facility. Both of these efforts very clearly follow the earthworks tradition of moving 'outside the gallery' and 'into the land' (Kuebler-Wolf 2013), in yet another return to the avant-garde motifs of rejecting art institutions and standardized frameworks of interpretation (Bürger 1994). *City*, however, is notable in this regard for its curated remembrances of ancient monuments, allusions to native mysticism and attempts to imagine the space of human activity through a time capsule, as it were. With its endlessly deferred completion mentioned in every occasional attempt to publicize the site, *City* has become a bona fide ruin in the making. Less a work of art as 'a work-in-progress' (Heizer 2014), this site's public availability will only happen, if it does happen at all, according to the express wishes of the artist. Barring exhibits or visitations of any kind, Heizer, in partnership with the Dia Art Foundation and others, has managed to finance the project independently during the course of five distinct phases of development. In fact, though *City* may appear to be a development project, which it is, the site has also managed to secure protection under Nevada conservation law (Heddaya 2014). Indeed with each expansion of the work, the newer structures establish deepening points of contrast and con-juncture with the surrounding environment. While in earlier phases Heizer focused on building massive concrete structures or buildings, the newer phases have become 'more pneumatic', resulting in 'raked dirt formations resembling hills, valleys and mountains' (Kimmelman 2005). Combined, these various sites give the impression of a total environment, an urban space that has been deli-cately arranged in the absence of any desire to be encountered, discovered, or contested. With his attachment to the aesthetics of the monument, the artist has voiced hopes that *City* will be admired both in its absence and erasure. For, as the site is composed from a series of shapes that 'align visually to make a frame around the structure' (Kimmelman 2005), it creates distinctions and relations that contest or interrogate the boundary between inside and outside, distinction and erasure.

The uniqueness of *City* is owed in part to the way that it contravenes the his-torically situated relationship between earthworks sculpture and questions of the urban; in fact, *City* challenges the assumption that earthworks sculpture should be classified as works of art at all. Because the site in question has been carefully protected from public view since construction began, Heizer has successfully challenged the long-held assumption that earthworks are made with the desire to stage encounters for a viewing subject or audience. In the absence of such a desire, I argue that *City* has acquired an element of *cityness* that paradoxically results from the site's abandonment, not from something that has taken place in spite of it. In other words, the consequences of this reversal are profound if we consider urban abandonment in general terms, beyond its artistic application or potential. Though the design of Heizer's site is premised on relations with the natural environment through a process of *techne*, these relationships serve the greater purpose of approximating ritual and ceremonial practices that are inherent to the social fabric of urban space. In effect, the cityness of *City* is determined by Heizer's musealization of the urban, which among other things acknowledges generational ties to the land. Now, because of this attempted musealization, *City* in turn cannot be determined by the criteria established by Henri Lefebvre (1991), for whom the spatiality of the urban constituted a privileged site for the production of the social, including the specific cleavages or tensions that accompany that production. To put it another way, I argue that *City* poses a methodological problem for Lefebvre's analysis of urban study, as abandonment according to that model has to be premised on the destruction of the social, and this destruction if anything becomes a design feature that Heizer exploits. Granted, the tradition of urban study that relies upon the spatial dialectic that originated with Lefebvre is crucial to the analysis of cities in general. As Edward Soja argues, for instance, by defining cityspace as a built environment, we tend to concentrate only 'on the distilled material forms of urban spatiality, too often leaving aside its more dynamic, generative, developmental, and explanatory qualities' (2000: 9). *City* not only fails to abide to these criteria, but it also exceeds the formalist and reductive alternative, whereby cityspace is reduced or reducible to a set of 'measureable and mappable configurations' (Soja 2000: 10).

Situating abandonment as parallel to the socio-spatial dialectic that Edward Soja (2000, 2011) proposes throughout his work, I suggest that Heizer's *City* is defined in fact by considerable activity. In spite of its desertedness, it is a location defined primarily by the creativity and originality of the artist, and, more concretely, by urban development. It is true, for instance, that *City* could not be understood exclusively as a reflection of Heizer's antagonistic relationship with the New York art scene, including his ongoing claim of being misrepresented and misappropriated by his peers. For Kimmelman (2005), Heizer's project is 'propelled by anger and resentment and monomania but also by Eros'. Though it contains the vitriol of Heizer's severed relationships, it also includes the seductive notion of 'sculpture as voluptuous, unspoiled and ecstatic' (Kimmelman 2005). Crucial to this vision is Heizer's generational memory of the land. As Kimmelman (2005) has observed, Heizer frequently describes the Nevada desert 'as virgin land', in the sense that 'he obsesses about the originality of his conception, about protecting his property and his art from violation by the rail, from developers hunting for underground water', and so on. In fact, Heizer developed a theory of 'cumulative observation' (Kimmelman 1999) in which to illustrate his resolute protectionism. Having inscribed optical illusions and permanent mirages into the monochromatic *lieux de mémoire*, he says that one must 'walk around [the sculpture], climb over it and later put it together in your mind' (Kimmelman 1999). In this, we are no longer dealing with 'the old convenient art object' (Kimmelman 1999). On the other hand, the urgency and necessity with which Heizer describes this interpretive framework is one that also corresponds to the impossibility of its execution. It is limited above all by a single gesture that both extends and limits the gaze of the onlooker. New methods have been devised to contravene this requirement. It has become well known, for example, that Heizer's paradoxical demands have been maligned by unforeseen circumstances at least since 2005, when Google Earth made satellite images of the work publicly available. Earlier images dating back to 1986 belong to the United States Geological Survey (Tarasen 2014). These mappable designs have indeed produced a representation of *City* that is capable of previously unimaginable mechanisms of specular control, having reoriented, challenged, and even disregarded the specific hermeneutic operation that Heizer insisted upon years ago.

The satellite images of *City* connect the work to a larger perspectival shift that Heizer could not have predicted at the time of this work's auspicious beginning. The viewer has once again returned to the frame, but this time with significant powers of exploitation, having acquired a vantage point that is situated high above. Anathema to cumulative observation, these images confound the artist's method of observation by visualizing *City* in its perpetual state of unfinishedness, while revealing its precise location in the desert for everyone to see. In a sense, the satellite images of *City* transform the site into an urban formation just like any other, falling victim as it does to the fluctuations of entropy, change, and surveillance. Connected to this epistemic shift is the sheer accessibility of a vantage point that has otherwise redefined the world picture by enhancing our capacities of attunement, having done so in ways that also risk the onset of incoherence and amnesia. In other words, by representing empty space through the interface of the satellite image, there is significant risk of replicating Walter Benjamin's famous pronouncement of a homogenous and empty time (2007: 262).

The epistemic criterion of these images has been lucidly described in attempts to historicize diverse cultural associations of 'public vertical space' (Parks 2013). In a recent article on this subject, Parks refers to the geopolitical challenges of keeping orbital space in the public domain from the time it was discovered back in 1927. Though an international treaty was eventually signed 40 years later with the aim of securing the orbital context for 'international peace, cooperation and collaboration', it has from the outset become a mere extension of the territorial divisions that line the surface of the planet below (Parks 2013: 65). Not surprisingly, orbital space today has become a playground for military infrastructures, telecommunications, and other corporate interests. For Parks, these competing actors on the orbital stage are implicated in a larger history that is itself 'firmly grounded in terrestrial politics' (Parks 2013: 68). Despite any repeated connection to the universalisms that I described in my introduction, Parks acknowledges that '"outer space" became a new historical, geographic, and theatrical/performative stage for shaping a discourse about rights and responsibilities, war and peace, security and risk' (2013: 66). With the popularity of Google Earth and the more recent efforts to make orbital space available to the public, the strategic elements that Parks describes have only been extended to serve the desire of private corporations, as the 'satellitization' (2013: 77) of the web has been directly tied to consumption patterns and profit margins. Acknowledging the terrestrial and political component is important here, in appreciating both the critical potential of the mappable arts as well as its limitations. In other words, by occupying the Archimedean point, as it were, the satellite image that represents vertical space for the public gives an impression of the urban that is remarkably abstracted from city life. *City*, on the other hand, performs a kind of double duty in this case. For, by reflecting the emptiness of space that results from the application of these technologies, the abandonment of *City* that is pictured by the orbital lens invites us to reconsider how any such urban space is socially produced and historically situated. Just like *Double Negative*, the work of *City* confronts the social production of space both in its absence and its erasure. As Kurgan writes, because maps not only represent space but *construct* spatial environments, because maps have become critical infrastructures in which we come to *know* the world, it is 'through a certain intimacy with these technologies – an encounter with their opacities, their assumptions, their intended aims – [that] can we begin to assess their full ethical and political stakes' (Kurgan 2013: 14). In this sense, a detour through mappable space allows us to assess the relation between urban abandonment and the artistic movement from which *City* came.

Gunkanjima Island and the archive of consumption

The ethical and political stakes that are raised by *City* are just as significant in the desert as they are in the sea. The image in Figure 11.3, for instance, belongs to an early collection by the Japanese photographer Ikkō Narahara. Entitled 'The sea wall; island without green', it appeared in Narahara's first solo exhibit which was devoted to themes of 'the human land'. The photograph depicts Gunkanjima Island, otherwise known as 'Hashima' or 'Battleship Island' for its distinctive shape, which is located just south off the coast of Nagasaki. In 1956 when this photograph was taken, Gunkanjima Island had one of the most highly concentrated populations in modern history, with more than 5,000 inhabitants on a stretch of land not much bigger than a football field. Abandoned now for more than 40 years, the island has become better known since 2009 when it entered the global circuit of the ruin obsession. Beyond this decontextualization of the site for the consumption of *memento mori*, I suggest that Narahara's photograph can be used as a mnemonic device in which to better situate the island in a larger history of Japanese colonial power, and, through this history, to develop a rationale for the insistence that its artefactual remainders be preserved in their current state. According to a study (Burke-Gaffney 2002), Gunkanjima Island during the Meiji period was targeted, along with neighbouring islands, for its reservoir of coal. During this period in the 1890s, the Mitsubishi Company entered the scene with new technologies that dramatically increased the production of natural resources. By the Second World War, the Japanese Empire was very active on

Figure 11.3 Ikkō Narahara (奈良原 一高) (1956) the sea wall; island without green
(© Narahara Ikkō Archives).

the island, having conscripted Chinese and Korean labourers to work in its mines, many to their deaths. The number of Japanese inhabitants increased after the war, as the island became a prime destination for unskilled workers. By the 1970s, however, coal production had lost ground to that of cheaper and cleaner fuels, leading the Mitsubishi Company to pull their investment in 1974 and force an immediate departure of its residents. The efficiency of this departure and the exquisite remains that were left behind is one important reason for the interest of memory tourists in the island today. In an attempt to respond – or rather convoke – this interest, the Japanese government in 2009 made a bid for the island to be included on the UNESCO list of World Heritage sites, arguing that Gunkanjima Island is home to one of the first-ever residential structures to be made from reinforced concrete (Yoshida 2009). Now, the Koreans in particular have made attempts to block the UNESCO bid, associating the site as they do with tragedy and exploitation by a colonial regime (Jiji Press 2014). Despite this gesture, the bid has already led to island renovations to prepare for international tourists, and to a marketing campaign responding to the intense scrutiny of the island following its debut in the popular James Bond film *Skyfall* (2012). Indeed, it was following the release of this film that companies like Google Earth descended upon Gunkanjima Island with the aim of preserving its urban remains from the onset of destruction by ocean winds.

Though the images of Gunkanjima Island in its abandoned state are noteworthy, the agenda put forward by Google Earth is rather different from that of the independent photographers who preceded them. In many ways, the visual archive of Google Earth results in a recolonization of space that, in turn, alters the particular histories to which they are associated. I turn to Ann Stoler's (2009) examination of the colonial archive as a way of situating the Google Earth initiative within a larger critical historiography. Stoler describes the mapping of colonial space as a discursive practice of ownership and authentication. The map, for Stoler, is a colonial archive that inscribes an unequivocal status of ownership onto faraway lands. Stoler, in fact, makes an oblique reference here to Jacques Derrida's (1988) concept of the signature as a way to illustrate the conjuncture between ownership and sovereign power. With watermarking, for example, the sovereign is able to define mappable space through a process of inscription, as through inscription the identity of a given space is authenticated as the sovereign's own. Stoler, however, argues that a further piece of the puzzle needs to be accounted for. She argues that while acts of authentication identify sovereign power, they also provide evidence of its limitations. Though counterfeit versions of the maps in question would be nullified by the watermark, for instance, Stoler herself insists that 'another sense of counterfeit does the opposite' (2009: 8), a sense in which the illegitimate version of the map poses an ongoing challenge to the sovereign that defines the other.

Given these conditions, the colonial archive serves 'both as a corpus of writing and as a force field that animates political energies' (Stoler 2009: 22). In fact, this suggestion contravenes Diana Taylor's (2003) theory of the archive and the repertoire. Taylor writes: 'it is difficult to think about embodied practice within the epistemic systems developed in Western thought, where writing has become the guarantor of existence' (2003: xix). This proposition is contradictory to Stoler's claim, because for Stoler it is the guarantor itself that magnifies the 'epistemological worries' (2009: 3) that define embodied practice in the first place. In other words, beyond the division that Taylor proposes between the (written) archive and (embodied) repertoire, the colonial archive that Stoler describes is one that already contains a repertoire in the shape of 'failed projects, delusional imaginings [and] equivocal explanations' (Stoler 2009: 21). Valuable as living documents that are complexly articulated between truth and the imagination, the sphere of writing gives us unique access with an outside world that returns us to the facticity of the archival document or image. If we then compare this investigation with the Google Earth initiative, I suggest that mapping is designed above all to domesticate abandoned space, abiding as it does in this case to the economic imperatives of transforming outmoded industrial space into a site designed for public consumption.

The narrative of colonial exploitation for which Gunkanjima Island had been known, at least by Koreans, has altered significantly since the airing of *Skyfall*. The disavowal of this past crystallized in June 2013 when the Google Corporation released Street View images of the film's location. Their release statement read, 'today you can explore the ruins of Hashima Island by pretending that you're James Bond waiting to be rescued' (Google Maps 2013). I argue that the popularization of the site by the Bond narrative has made the Island's history subject to a double articulation. The first articulation frames the ruins of Gunkan-jima Island as a haunting display of early industrialization, evoking the apocalyptic certainty that lies at the heart of the modern world. The remoteness of this site only further capitulates the global reach of this imminent catastrophe. The second articulation is conceived as an effort to salvage the island's remains, which builds on the archival function of visual media as a tool of preservation. The archival responsibility that Google Earth has assumed in recent years has led to some noteworthy projects, such as the 2014 effort to chart the impact of climate change on polar bear migrations in Canada's North (Bogo 2014). Now, while Google Earth extends their archival responsibility to the Gunkanjima project, they simultaneously depart from the seriousness that such work demands. Offering a portal to the imagination, as it were, the Gunkanjima archival project is used in the service of Google's desire to become an irreplaceable supplement of entertainment media. In a groundbreaking essay from 2009, the geographer Paul Kingsbury suggests that we move beyond defining geospatial media with exclusive reference to that of 'surveillance, warfare, and invasion of privacy' (Klinkenberg in Kingsbury and Jones 2009: 503). If Google Earth is substituted for film, in other words, it may be possible to go beyond this realm to that of another, more affective channel in which the images they produce can be brought together with our living world – a movement, Kingsbury writes, towards 'the un-tethered openness of Dionysian uncertainty' (Kingsbury and Jones 2009: 503). Now, Google Earth supplies many precedents for Kingsbury's argument: from the mapping of Harry Potter's Diagon Alley and the *Game of Thrones*, to locating your favourite album covers or comparing street scenes from Victorian-era paintings, there are indeed 'new readings of techno-culture that are far from the dystopic options of Apollonian control' (Kingsbury and Jones 2009: 505). Street View images of Gunkanjima Island are particularly revealing because they occupy what Steven Shaviro (2010) describes as the realm of 'post-cinematic' media, in which the cinematic arts are subsumed by a collection of diverse visual media.

I argue, however, that while Google Earth represents an extension of mappable space beyond the cinema, it does so in a way that uses cinema paradoxically to communicate this reflexive capability. For instance, the archival remains of Gunkanjima Island are made comprehensible only by the detour of cinematic narratives that brought it to our attention in the first place. In *Skyfall*, the Island is briefly featured together with a studio reconstruction of its urban interiors. It is introduced at a crucial transition point in the narrative, marking the first explosive encounter between Silva, the villain played by Javier Barden, and Bond, played by Daniel Craig. The island is cast as a strategic hub for Silva's illegal activities, which includes stock market fraud, rigged elections, and terrorist bombings. In a dramatic confrontation between the foes, Silva (who has captured Bond) refers to his most beloved emotional and symbolic attachments – England, the empire, and M16 – suggesting to Bond that he, too, is 'living in a ruin' but 'doesn't know it yet'. The larger implication of this exchange is whether the Bond franchise itself can validate its relevance for a new generation of viewers, suggesting that by going back to the roots of the Bond mythology, as *Skyfall* does, one can reimagine the future potential of the series. Indeed, from this vantage point of critical reflection, the Island's ruins become an allegory of the desire for rejuvenation. On the other hand, there appears to be an effort here to establish an equivalence between London and Gunkanjima Island as opposing sites of agency and technological power, in a way that confirms Tom Conley's suggestion that the cinema shares 'an implicit relation with cartography' (2007: 1). For instance, Conley writes that a film's 'locational imaging' is often implicated in 'bilocation' (2007: 3), suggesting that in the presentation of discernible markers of location on the screen, a film will seek to occupy a space between the real and the imagined. For example, in the scene immediately following Bond's safe return from the island, we find that despite having captured Silva and returned him to London for an arbitration, Silva has managed to infect the electronic systems of the nearby espionage command centre with a computer virus. Illustrated by blood-red striations that splash across the centre's main screen in a chaotic presentation, we soon discover a narrative itinerary. As the striations shape into locational markers, the virus mutates and spreads while the image transforms into an impressionistic version of London's underground tunnels, which in turn provides important clues for the viewer regarding the whereabouts of the next action-packed encounter.

The life of the Street View *flâneur* is anything but enchanting, as the world it depicts could not be further from the action-packed universe of the James Bond film. Street View participants are abandoned to the logic of the interface and tasked with visiting a world that is poor and empty by comparison. The recordings of Gunkanjima Island only bring that emptiness into view by signalling the threat of destruction that lies beneath the surface of every human community.

Figure 11.4 Street View capture of Gunkanjima Island (© 2013 Google).

In a sense, the truth of that threat is most strongly expressed in scenes of urban abandonment, where the human community in question has already been reduced to its garbage. Now, the point-of-view navigation of the Street View device has the benefit of preventing the geospatial detritus from being aestheticized on terms that are not its own. The Street View interface differs radically from the photographic work of the Canadian Jon Rafman, for instance, which features screen captures of Street View scenes that evoke a Barthesian punctum to signal greater depth. Paraded in art galleries around the world, Rafman's project may be valuable for encouraging a social commentary around these technologies, but it also fundamentally contravenes the labour of solitude that the Street View experience demands. Returning to Marc Augé's pronouncement, the labour of solitude that defines that experience is headed above all by an individual in established relations to the non-place, in which 'solitude is experienced as an overburdening or emptying of individuality' (Augé 2009: 87) that must in turn be experienced if a future is to be imagined at all.

12 Waste and value in urban transformation

Reflections on a post-industrial 'wasteland' in Manchester

Brian Rosa

Introduction: place-as-waste

This chapter examines changing conceptions of wastelands, particularly in the case of former industrial districts of cities in advanced capitalist economies. It explores the historical development of the wasteland concept, as well as informal and formal reappropriations of 'wasted' post-industrial spaces, in processes of urban social, cultural, economic, and ecological change. In these ambivalent (Jorgensen and Tylecote 2007), vague (Solà-Morales Rubió 1995; Miller 2006; Barron 2014), and sometimes contemptuous (Armstrong 2006) sites, what is wasted in an urban wasteland, and to whom is this a problem? In exploring the discursive construction of place-as-waste, the dialectical relationship between waste and value becomes of central concern.

Among the most sustained considerations of so-called 'urban wastelands', and the process of wasting more broadly, has been the urban planning and design theorist Kevin Lynch (1960, 1972; Appleyard *et al.* 1964), particularly his final, posthumously published book, *Wasting Away* (1990). In this exploratory text, Lynch departs from prescriptiveness of 'good city form' and urban 'imageability' to appeal for the acceptance of wasting as a necessary social, ecological, and material process. To him, waste was:

> what is worthless or un-used for human purpose. It is a lessening of something without useful result; it is loss and abandonment, decline, separation and death. It is the spent and valueless material left after some act of production or consumption, but can also refer to any used thing: garbage, trash, litter, junk, impurity and dirt. There are waste things, waste lands, waste time and wasted lives.
>
> (1990: 146)

Lynch was primarily concerned with the 'wasting of place' in universal processes of wasting and decay, and specifically with how planning could accommodate the process of 'wasting well'. While there are limitations to normalization of waste and the prevailing political-economic relations that produce it, this process-based approach helps contextualize wastelands with the social production of space (Lefebvre 2009 [1974]). The quality of waste is defined by circulation (Trotter 1988), and wastelands are a product of changing flows of people over ever-increasing distances, aided by fixed assets embodied in transportation and communication infrastructures (Harvey 1982; Graham and Marvin 2001). Understanding an urban wasteland means interpreting the changes in capital flows and shifting power geometries (Massey 2005) at global and local scales: rather than circulation of waste, this shifts attention to waste as a result of changing patterns and scales of circulation. Drawing from critical urban theory (Brenner 2012), this analysis requires a non-instrumental analysis of urban change that focuses on conflict rather than consensus.

This chapter seeks to explore how the flight of industrial capital from cities has generated the production and perception of urban wastelands. Proliferating in the *terrains vagues* (Solà-Morales Rubió 1995) of deindustrialization and fragmented spaces of infrastructure, wastelands are defined by their disorderly or unmaintained appearance, their functional or perceived underutilization, the anxiety they inspire (Picon 2000), and their economic underperformance (Di Palma 2014). Therefore, their wastefulness is defined, first, through an apparent lack of social or economic value and second, through the presence of waste matter, ruination, transgressive social behaviour, and 'wild' nature. As temporal as they are spatial (Stavrides 2014), wastelands demonstrate that waste is not only 'matter out of place' (Douglas 2002 [1966]) but, at a particular moment, places that do not matter. However, as these sites generally go overlooked and unmentioned, their classification as wastelands often occurs at the moment when development pressure makes reconfiguration profitable.

Wastelands have held an ambivalent position in the post-industrial urban imaginary (Campkin 2013): to urban planners, they may be a 'valuable strategic asset for localities' (Bowman and Pagano 2004), with the emphasis of increasing intensity of use and property tax revenues. With public subsidy and increasing policy emphasis on urban densification (DETR 1999), these sites have become the most important locations for contemporary urban redevelopment (Berens 2011). However, imperatives for redevelopment and increased urban density are complicated by the fact that wastelands are being reimagined as unique habitats of urban biodiversity and aesthetic curiosity (Gandy 2013) and 'loose' public spaces (Franck and Stevens 2006), often defined by the intermingling presence of industrial ruins (Edensor 2005; Garrett 2013). Clearly, there are multiple *values* embedded in waste spaces (Gidwani and Reddy 2011), and economistic concerns for the valuelessness of urban wastelands often subsume social and cultural values (recreation, heritage, public space, shelter) and ecological value (which may be justified by cultural or economical values, i.e. 'ecosystem services') against economic value (embodied in exchange value of land and the potential profits of redevelopment).

To illustrate the conflicting values embedded in an urban wasteland, I focus on the former docklands of Pomona Island in Manchester, England. I aim to distinguish different discourses mobilized by various interest groups in relation to the current and future use of this superficially 'empty' site. This draws on ethnographic fieldwork I conducted between 2009 and 2015, semi-structured interviews conducted between 2009 and 2015, and analysis of planning and policy documents from the 1980s to present. As I will argue, the mobilization of the term 'wasteland' in public discourse is a normative instrument, utilized to justify the reconfiguration and profit-making potential of a site and to de-emphasize any values that conflict with this goal. In short, it is a term that obfuscates as much as it reveals.

Waste + land: landscapes and land uses in a historical context

In terms of thinking of land as a category of waste, it helps to consider the mutually constitutive relationship between waste and value in classical political economy. Locke (1988 [1681]) saw the transformation of waste to usefulness (in the case of land, through cultivation or other 'improvements') as the defining moment of political modernity. Gidwani and Reddy (2011) argue that waste is 'indexical of the necessity for an ordering rule of property' (Gidwani and Reddy 2011: 1626) and that the concept of 'waste' is 'the specter that haunts the modern notion of "value"' (Gidwani and Reddy 2011: 1627), since it emphasizes inefficiencies, insufficient wealth generation, and unexploited resources. Both 'waste' and 'value' also imply moralized connotations and economic quantifications of wealth (or lack thereof), with 'value' performing as both a measure of economic output and a moral virtue of conduct. However, there is also a central contradiction to thinking of land as something tradable: it is, according to Polanyi, a 'commodity fiction' that is 'an element of nature inextricably interwoven with man's [*sic*] institutions' (2001 [1944]: 187), making its isolation and marketization 'perhaps the weirdest of all the undertakings of our ancestors'. In English agrarian history, the enclosure movement (Polanyi 2001: Chapter 3), offers insight into the contemporary image of, and discourse surrounding, urban wastelands.

The expropriation of commonly owned and managed pasture, or commons – famously pointed to by Marx (1990 [1867]) as an exemplar of 'primitive accumulation' – was largely justified through elites' claims that commons were wasteful due to their economic under-productivity. With the gradual enclosure of the commons, landed elites marked open fields and commonly held lands as waste ground, and by the early nineteenth century wastelands and commons were increasingly being characterized as mutually constitutive and interchangeable (Goldstein 2013). It was from this context that the term 'waste land', later 'wasteland', emerged. Di Palma (2014) points to changing land uses and landscape ideals in eighteenth-century England to emphasize that the visual orderliness of land became indexical to its appropriate management: wasteland was also becoming a recognized aesthetic category. To Di Palma, wastelands illustrate how 'anti-picturesque' landscapes have influenced shifting conceptions of beauty, sublimity, and the moralized economies of 'improving' uncultivated or common lands. In this sense, a wasteland is,

> united not by what it is or what it has, but rather, by their absences.... The emptiness that is the core characteristic of the wasteland is also what gives

the term its malleability, its potential for abstraction; a vacant shell, it lies ready to include all those kinds of places that are defined in negative terms.

(Di Palma 2014: 3–4)

In this regard, the idea of the wasteland has long served as a useful rhetorical device for elites to present the rationalization and marketization of insufficiently productive land as a virtuous endeavour.

Waste – in terms of land use as well as environmental degradation – re-emerged as a central motif of nineteenth-century industrial urbanization, with Manchester as an archetypal 'shock city' (Platt 2005) illustrating the excesses of industrial urbanization (Mumford 1961; Hall 1998). As the first industrial city, Manchester was the subject of considerable political debate regarding emerging concerns of industrial waste and value. Waste was among the central foci of numerous international visitors to the city in the 1840s, including Tocqueville (1956 [1845]: 105), Faucher (1969 [1844]), and Engels (2009 [1845]). Tocqueville pinpointed the moral ambiguity of the city and its perceived wasteful-ness: 'From this foul drain, the greatest stream of human industry flows out to fertilize the world. From this filthy sewer pure gold flows' (1956: 105). Through industrial urbanization, wastelands proliferated as spatial by-products of the creative destruction of urban modernity (Berman 1988; Dennis 2008). To contemporaries, the waste of laissez-faire industrial urbanization was typically normalized, if regretfully, as an inevitable spatio-material by-product of economic progress (Joyce 2003). The externalities of industrialization would be tolerated in the name of progress with the mitigation of excesses and the separation of land uses leading to the professionalization of city planning, civil engineering, surveying, and other specialized fields focused on apportionment and management of land use (Hall 1998).

The rapid industrialization of Manchester, like other cities that followed (see Cronon 1991; Platt 2005), was enabled by coinciding revolutions in transportation infrastructure, which allowed for the colonization of north-west England's countryside and its connection to global circuits of trade. These same revolutions in transportation, communication, and the exploitation of cheap labour that established Manchester's industrial dominance played a major role on the city's deindustrialization since the Second World War, exacerbated by the decline of imperial trade relations as well as the decentralization of industry and housing ushered in by the automobile.

By the 1960s, industrial decay and abandoned terraced housing offered a mnemonic device for the decline of a way of life in northern England (Taylor *et al.* 1996; Crinson 2005). The results of globalization and decentralization on the urban fabric were proliferating swathes of wastelands encircling the commercial cores of cities, visually signalling economic neglect and an indeterminate future for (re)use (see HM Stationery Office 1963; Civic Trust 1964; RSA 1965; Barr 1969). Attitudes toward the disorderliness of industrial landscapes shifted from ambivalence to disdain as they became decreasingly productive (Barr 1969). An organizing motif of concerns about waste and deindustrialization has been dereliction, the highly visual abandonment and dilapidation signalling the flight of industry and investment. An especially 'wicked problem' (Rittel and Webber 1973) for urban planners has been 'brownfield' sites: tracts of land whose previous industrial use has contaminated soil to the extent that redevelopment requires intensive remediation, often with considerable public subsidy. Nabarro (1980) identified three specific types of post-industrial urban wastelands in Great Britain, differentiated by the reasons leading to a site's disuse. These included land left over from slum clearance and urban renewal schemes, disused former industrial land, and sites left vacant by speculative landholders waiting for the moment when selling or redeveloping these sites would be profitable. While the phenomenon of speculative landholding was still in its nascent stages in the early 1980s, Nabarro's explanatory theory of urban wastelands offered a cautionary note: what appears superficially to be an abandoned plot may actually be the subject of significant 'hope value' on the part of investors (Ball *et al.* 1998: 34).

Property speculation has, indeed, become one of the central targets of Marxian geographical political economy since the 1970s: with private landowners treating real estate as a secondary circuit of capital, they have incentive to take advantage of the crisis of uneven development (Smith 1984) and place-specific devaluation through purchasing land as a form of investment. This, in turn, often aids or leads to widespread devaluation of fixed capital in inner city locales (Merrifield 1993). Through uneven development,

> capital attempts to seesaw from a developed to an underdeveloped area, then at a later point back to the first area which is by now underdeveloped.... Capital seeks not equilibrium built into the landscape but one that is viable precisely in its ability to jump landscapes in a systematic way.
>
> (Smith 1984: 198)

In this sense, an urban wasteland may be a frontier for the realization of the 'rent gap': after prolonged neglect, capital is attracted to urban land at the peak of its devaluation, when the difference between capitalized ground rent and potential ground rent becomes sufficient to redevelop a site in a 'higher' use. The rent gap theory offers limited explanatory function, but it effectively captures the spatio-temporal aspects of urban decline and renewal to illustrate the cyclical nature of capitalist investment and disinvestment. Reflective of Polanyi's 'commodity fiction' (2001), land is one such investment that can increase in value without any improvements being made or any productive use. For this reason, a wasteland may superficially appear abandoned, but may in fact be the subject of significant economic interest, as will later be demonstrated in the case of Pomona.

With speculative landholding of post-industrial sites, waste signals a different sort of underutilization and under-productivity: land and structures exist as an appreciating investment without their owners' maintaining any significant use value. Among the most common examples are surface-level car parks. In some cases, any commercial use may be avoided, as interim uses may complicate future plans for a site. This behaviour turns the idea of waste as economic under-productivity on its head. This is not to say, however, that 'empty spaces' proliferate in contemporary cities. When the use and purpose of a site is indeterminate or vague and its owners are unknown or unacknowledged, it is common for members of the public (most likely proximate communities) to animate these sites with myriad informal uses (Groth and Corijn 2005; Carney and Miller 2009; Sheridan 2012). In other words, urban wastelands are often reappropriated as communal spaces for gardening, recreation, and play. Depending on who appropriates these spaces and how (artists versus homeless people, for example), these common uses may be encouraged or discouraged, and may have varying impacts on the market valuation of urban sites (or entire districts).

Theorizing urban wasting in the post-industrial city

Reversing or halting the wasting of urban space, as normative planning and design theory tends to reinforce, requires adaptation, maintenance, and rationalization of wasted, vacant (Bowman and Pagano 2004), or 'lost' (Trancik 1986) spaces. Taking this perspective for granted, Lynch's reflections on the passage of time (1972) and waste (1990) are based on the assumption that 'changes, when managed, are meant to lead to more desirable states, or at least to avoid worse ones', that 'underlying change is either desirable or inevitable' and that 'the problem [of planning] is to deal effectively with the transition itself' (1972: 190). This outlook, emphasizing the inevitability of wasting and the need for urban repair, has become dominant in urban planning and design theory (see, for example, Southworth 2001; Berger 2006). Even if we are to accept urban wasting and capital mobility as a naturalized cyclical process, this still raises the question of whether the managed reconfigurations of wastelands produce 'more desirable states', for whom, and through what process.

Beyond understanding who benefits from the rationalization of urban wastelands, which is clearly situational, we must consider that paradigmatic notions of ideal post-industrial urban landscapes are also in a state of transition. For example, strategies for redevelopment based on industrial heritage – where the industrial built environment is considered a culturally and economically valuable amenity – demonstrate that industrial landscapes within themselves may be considered desirable, and therefore, marketable (see Zukin 1995). Just as waste is recycled, the recycling of the industrial built environment is a driving aesthetic sensibility of post-industrial redevelopment (Campkin 2013). What constitutes a desirable landscape is not static, neither in its physical form nor in the type of sociality it engenders.

One question that arises from this scenario is as follows: how should planning and design processes manage urban wasting, and to what extent are wasteland redevelopment schemes a reflection of democratic participation? This question is one of the driving motivations behind alternative theorizations of the value(s) of urban wastelands, from Lynch and beyond. After decades dominated by the wholesale erasure of urban wastelands within a modernist framework, designers and urban theorists since the 1980s have expressed ambivalence toward approaches that ignore the unique social values retained in supposedly wasted spaces (Solà-Morales Rubió 1995). The ideal uses of urban wastelands have arisen as a subject of considerable discussion in the design fields, particularly around their commonplace reappropriation as informal public spaces, as a sort of urban commons. A more emancipatory tone is evident in Cupers and Miessen's *Spaces of Uncertainty* (2002), where the authors promote indeterminate spaces in Berlin, Germany as an antidote to the overdetermined, homogeneous, tightly regulated, and increasingly privatized urban public realm (also see Carmona 2010; Minton 2009). The authors critically reflect on the desire of architects to conceptualize urban voids – open areas without clear function – as an opportunity for design practice and meaningful reintegration within the fabric of the city: spaces to be 'colonized' (Van Dijk 1996). Furthermore, Loures (2015) finds that in post-industrial landscapes, designed-based approaches tend to be 'primarily focused on aesthetics, leaving society's other main goals to secondary status' (72). There is a clear instrumentality in designers characterizing leftover urban spaces as 'voids' in need of reclamation, considering that the reprogramming of post-industrial spaces is a prime opportunity for promoting their professional practice. However, 'voids' clearly have value for those people who make use of them, whether through temporarily appropriation or longer-term squatting (Doron 2000). In this instance, the discursive connection to wastelands and commons is straightforward. Meanwhile, studies exploring perception of urban derelict land find that local residents accept leftover spaces as recreational areas and parklands, especially if they are accessible and are minimally maintained (Hofmann *et al.* 2012), suggesting that concerns over the appearance and unruliness of wastelands is foremost a preoccupation of designers and policymakers.

Looking beyond questions of human inhabitation, the (e)valuation of urban wastelands has become more complicated amidst increasing interest in urban biodiversity in planning, design, and policy. Precisely due to their neglect, leftover and liminal urban spaces have been demonstrated to foster much higher levels of species richness than traditional parks and public spaces (Rink 2009). This, in turn, has begun to influence urban policy, with urban biodiversity and green infrastructure (TEP 2008) considered an essential element in mitigation of, and adaptation to, climate change (Hall 2013). Likewise, in Britain, the conservation establishment is increasingly acknowledging the unique biodiversity of urban wastelands (Baines 2012: xiii) and their integral roles in providing 'green infrastructure'. Economists and planners have indeed quantified the 'ecosystem services' provided to cities by their wastelands due to their role in carbon sequestration (Robinson and Lundholm 2012), maintaining a financial valuation of land use while questioning the assumptions embedded in traditional approaches.

The increasing appreciation of urban nature reflects not only a concern for sustainable development, but changing aesthetic tastes and ecological sensibilities: as Di Palma (2014) notes, cultural and economic evaluations of ideal landscapes are dynamic and intertwined. This could be observed in England as early as the 1970s, when the untamed frontiers of wild nature had come to define

deindustrializing urban districts. British naturalist Richard Mabey (1973) celebrated the 'unofficial countryside' found on the fringes of London, noting that
'the natural world is indifferent to ... the clutter and ugliness ... of our urban
environments' (1973: 14). Mabey, and many of his naturalist followers, did not
so much celebrate the dereliction and neglect of wastelands, but exalted the
resilience of the flora and fauna that animated these sites. 'It is not the parks', he
noted, 'but the railway sidings that are thick with wildflowers' (1973: 12). More
recently, in *Edgelands* (Farley and Symmons Roberts 2011), two poets journey
into the 'true wilderness', romanticizing unnoticed, in-between spaces such as
gravel pits, landfills, and industrial parks along the urban fringes of north-west
England. Along these lines, there is an emerging romantic sentiment toward
post-industrial urban landscapes, with the affordances of disorderly and unmaintained urban spaces being increasingly celebrated (Edensor 2007, 2005; DeSilvey and Edensor 2013). Quite often this is based on the fascination of nature
'overtaking' the industrial built environment and the uncanny experience of
modern ruins, a trend deeply embedded within the rising profile of 'ruin porn'
(Millington 2013) and the subculture of urban exploration (Garrett 2013).

Jorgensen (2008) has mobilized the term 'urban wildscapes' to signify 'urban
spaces where natural as opposed to human agency appears to be shaping the
land, especially where there is spontaneous growth of vegetation through natural
succession' (Jorgensen 2012: 1). Sheridan defines wildscapes as 'any area,
space, or building where the city's normal forces of control have not shaped how
we perceive, use, and occupy them' (2012: 201). Thus, the wildness in wildscapes refers to disorderliness both in terms of non-human and human appropriation of urban spaces that are seemingly outside 'normal forces of control', land
use regulation, traditional forms of maintenance, and surveillance. We can see
clear overlaps in the celebratory discourse on the socio-ecological reading of
urban wildscapes and exaltation of underdetermined, 'loose spaces' (Franck and
Stevens 2006) and 'spaces of uncertainty' (Cupers and Miessen 2002). The
increasingly celebratory cultural attitude toward the naturalistic and aesthetic
affordances offered by urban wastelands and intermingling industrial ruins helps
to explain a gradual transition through which they are actively incorporated in
contemporary landscape design aesthetics dominated by adaptive reuse and
urban greening.

Clearly, regardless of how urban wastelands may be represented in planners'
maps and redevelopment frameworks, they rarely exist as *tabula rasa* (Doron
2000). Beyond considering the merits of different design approaches to leftover
industrial space, we might ask a set of more critical questions: when and where
urban spaces are problematized as 'wasted', underutilized, or empty? How does
this relate to the logic of capital accumulation through the discursive shaping
of transitional urban sites? The following sections explore how one site in
Manchester can inform some of these questions about waste and value in the
contemporary post-industrial city. After establishing the relevant history of
Pomona, I will focus on its contemporary representations in relation to its past
and future uses.

Pomona: Manchester's infrastructural island

Manchester is often held up as a model of post-industrial urban renaissance (Peck and Ward 2002; Hebbert 2010), a transition often attributed to the town hall's strategic adoption of municipal entrepreneurialism (Harvey 1989a), through promotion of property-led redevelopment since the 1980s (Quilley 2002, 1999; Ward 2003; Allen 2007; Leary 2008). Still, as much as the city has experienced a boom in commercial, residential, and retail development from the 1990s to the financial crisis of 2008, the legacy of deindustrialization is still readily apparent on its fringes. Perhaps the most prominent of Manchester's wastelands is Pomona Island, straddling the south-west corner of Manchester at its border with Trafford and Salford. At over 20 hectares, Pomona remains one of the largest undeveloped sites in close proximity to the city centre, and one of the largest green spaces in the city. It has existed for decades as a 'dead zone' (Doron 2007b) on which new planning aspirations have been projected, though at face value it has remained a relatively 'empty' space.

Pomona is a residual space of considerable scale: it is completely delineated and dominated by the transport infrastructures. Even its status as an 'island' reflects successive layering of transport infrastructure: rather than being a geomorphic island, Pomona is an anthropogenic space more reminiscent of novelist J.G. Ballard's (1973) *Concrete Island*. The site's planning boundaries are defined by the Metrolink tram viaduct (1999) the Manchester Ship Canal (1894), and its Pomona Docks. The Bridgewater Canal (1761) also passes through the site, which is further cut off from Central Manchester by the Cornbrook railway

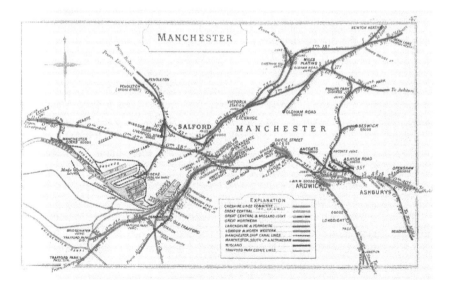

Figure 12.1 The infrastructural network of Manchester at the peak of its industrialization. Pomona Docks may be found just left and down from the centre of the image (image source: Railway Clearing House, 1910).

viaduct (1877) and the arterial (A56) Chester Road, both running parallel to the Metrolink viaduct. It attests to the 'splintering' effect of infrastructure (Graham and Marvin 2001) and the obduracy of infrastructural networks (Hommels 2005) that often serve to produce and define leftover urban spaces. Though the site's modern history has been defined by transport infrastructure, during the Victorian era it maintained the characteristics of a peripheral countryside, yet to be transformed into a space of transfer for the industrialized city. Pomona Gardens, which occupied the site from the 1830s, was a pleasure garden and orchard. By the 1860s, it had become the site of the Pomona Gardens Palace, an event centre that could accommodate more than 30,000 people. The Palace occupied the site until the anticipated completion of the Manchester Ship Canal led to the clustering of factories around the site in the 1880s, and it was ultimately shut down by an explosion at a chemical plant in 1887 (Flynn 2013). By that point, Pomona had fully been absorbed as a 'glocal' (Swyngedouw 2004) space of intermodal transport logistics: one need only look at the North American names of the Ship Canal's nearby basins (Ontario, Erie, and Huron) to consider the thoroughly global nature of this locale.

By the mid-twentieth century, the redundancy of the area's transport infrastructure was a testament to Manchester's slowing industrial metabolism. Between the 1960s and 1980s, the Bridgewater Canal and the Cornbrook Viaduct had become disused, and the Ship Canal quickly became obsolete due to its inability to accommodate new oceangoing container ships. By 1982 all of the Ship Canal's docks had been closed, with over 3,000 jobs lost (Salford City Council 2008).

For the following three decades, Pomona's future has been the matter of successive waves of speculation, interjected by prolonged periods of neglect. Upon the closure of the docks, the site was not completely abandoned, but overtaken by light industrial premises. Some of the docks were filled in, but the soil has remained deeply contaminated. Though disused lots on the site were increasing between 1982 and 1986 (Turner 1989), into the mid-1990s there were a variety of scrapyards, automotive repair businesses, and construction-related firms operating in the area (Conran Roche 1989). In the 1980s, in the era when municipal socialism still predominated local politics, Manchester City Council had earmarked this area as a site for industrial retention, and these plans had been slow to change due to a lack of development pressure.

However, aspirations to redevelop Pomona were heightened by the much-celebrated regeneration initiatives in the nearby district of Castlefield in the 1980s and 1990s. This former industrial district, also dominated by canal and railway infrastructure, was transformed by public- and private-sector actors into an Urban Heritage Park: a leisure and tourist destination focused on the consumption of industrial heritage (see Degen 2008; Leary 2011; Madgin 2010; Rosa 2014). Castlefield also became one of the sites where speculative capitalization on the rent gap (Smith 1984) could first be seen in Manchester, when local bookmaker Jim Ramsbottom began purchasing decaying warehouses for the price of salvaged brick and timber in 1982 (Parkinson-Bailey 2000: 289), with the intention of later developing the land. From 1988 to 1996, under the tenure of the Central Manchester Development Corporation (CMDC), Castle-field's development as a heritage tourism and recreational destination was pre-dicated on the displacement of 'low value' or 'bad neighbour' land uses – scrap yards, auto repair shops, timber merchants – to Pomona and adjacent Cornbrook. The continued existence of industrial usage of Castlefield conflicted with its new re-imaging as a musealized, post-industrial landscape.

The CMDC (see Figure 12.2) was an Urban Development Corporation desig-nated by the national government to encourage commercial expansion in the city's Southern Fringe: armed with significant public funds and the power of compulsory purchase, for eight years this quango was able to supersede Man-chester's city government in spearheading property-led development initiatives. Like Castlefield, Pomona was part of the CMDC's massive remit area, albeit the most peripheral and least invested site. Out of concerns raised by consultants to the CMDC that 'continuation of the existing uses would be likely to detract sig-nificantly from the ability of the former Docks area to attract new investment and businesses' (Symonds Travers Morgan 1996: 1), most of these industrial tenants were ultimately displaced and the site left almost completely vacant. As was common in the fringes of Manchester city centre under the tenure of the CMDC, the displacement of low-intensity industrial land uses – themselves cast as markers of wasteland – was a key element in a wave of 'environmental improvements' considered necessary to attract property-led redevelopment. For

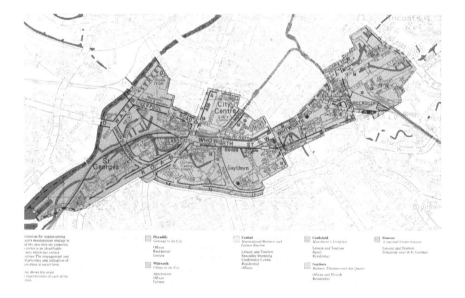

Figure 12.2 Map of the CMDC's remit area. Pomona is the area furthest to the bottom-left of the image surrounding the dock labelled 'No. 1 Dk.' (image source: CMDC, 1990).

this reason, throughout the 1990s, additional scrap metal recovery and other light industrial businesses were denied planning permission to operate nearby. Removing all use value of a site was considered an improvement over maintaining land uses that were perceived as jeopardizing future development. According to a report produced for the CDMC,

> Although comprehensive redevelopment schemes have been put forward in the past none have come to fruition partly as a result of the site's 'hidden' location, difficult access, potentially high infrastructure costs and because regeneration resources have tended to be focused elsewhere.
>
> (Symonds Travers Morgan 1996: 17)

By that point, planning in Manchester was increasingly shifting from a role of land use regulator and manager to an active entrepreneurial agent, channelling investment and speculation onto prioritized sites (Quilley 1999, 2002).

Considering contemporary debates regarding the use and meaning of the site, it is notable that plans from Manchester City Council, the CMDC, and the English Tourism Board (1989) throughout the 1980s and 1990s aimed to re-establish Pomona as a regional leisure park, building on the success of Castlefield and the area's history. The CDMC had proposed to 'explore ... the scope for creating a major landscaped area ... which would capitalize upon the waterway links' (1989: 4.16), to create a 'vital green lung close to the city centre' (6.13). An ambitious document by the design firm BDP called the *Waterways Guide*, commissioned by the CMDC, envisioned a 'contemporary sculpture park with commercial leisure activity', including 'an extensive open area called City Park' (BDP 1989). Some infrastructure for this park was installed in the late

Pomona –
A Regional
Leisure Resource

There are two distinct parts to this sector: St. Georges and Pomona Docks. St. Georges is a run down mixed industrial area which requires environmental and infrastructure improvement. Pomona is a largely derelict dockland area, linking the Castlefield area to Trafford Wharfside and Salford Quays. Only part of the former docks is within the Development Corporation's area, the rest lies within the area of Trafford Park Development Corporation. The area is almost completely surrounded by water and currently has major accessibility problems. This site has the potential to become a major regional leisure park at the heart of the conurbation.

Figure 12.3 The newly installed landscape promenade along the banks of the Manchester Ship Canal at Pomona (image source: Development Strategy for Pomona. CMDC, 1990).

1980s, including a landscaped promenade along the Ship Canal including decorative lamps, benches, railings, and planters (see Figure 12.3), but its public access never came to fruition. The early 1990s saw few additional changes to the site. In 1993, most of the land at Pomona was passed on from the city of Manchester to Trafford, though the CMDC retained responsibility for the Pomona site (Symonds Travers Morgan 1996). By 1994 plans for Pomona had been shelved, with reports noting that 'the area remains much as it was at the beginning of the period' (Deas *et al.* 1999: 222). With the CMDC scheduled to be dissolved in 1996, the future of the site was unclear. Pomona was the district that had clearly received the least attention by the CMDC (Kitchen 1997: 141), to the point that it was entirely left out from their report chronicling their achievements (CMDC 1996).

The transfer of ownership of most of the Pomona site to Trafford was more significant than it might initially appear: the council of Trafford is heavily influenced by The Peel Group (formerly Peel Holdings), an infrastructure and property investment conglomerate with assets in excess of £5 billion headquartered there (ExUrbe 2013). Since 1987, Peel had become the primary landholder of the Pomona site, owning all but one small site operated by a scrap metal recycler, having acquired the properties of the former Manchester Ship Canal Company. Owning 15,000 hectares of land and water in the UK, with a portfolio valued at £2.3 billion (The Peel Group 2015), Peel is one of the largest privately owned property companies in the United Kingdom (Harper 2013), owning or developing nearly all canals and much of the canal-side property in greater Manchester. In this sense, the stalled public investment in the site and the indeterminacy of Pomona's future can be understood through exploring the property's ownership. Aspirations for creating parkland were quickly pushed aside. In interviews, local planners have suggested that Trafford would not be likely to express interest in creating a public space at Pomona, since its location would lead to use primarily by Manchester and Salford residents. None of this is at all clear to nearby residents or many of the site's users, of whom few are aware of council boundaries or, indeed, the fact that Pomona is no longer under the jurisdiction of Manchester City Council. However, it is far from a 'no man's land'. After all, as Peel describe their business model, 'our approach is primarily driven by planning and development opportunities – we retain assets as a complement to our longer-term strategic projects' (The Peel Group 2015: 4). Their company motto is 'determination, perseverance and patience'.

The construction of the Eccles line Metrolink tram extension, which passes through the site on a concrete viaduct and was completed in 1999, was anticipated to finally revalorize Pomona. With the island having been cut off from the rest of the city by transport infrastructure, another layer of infrastructure was expected to revive it. The line runs from central Manchester on a previously disused Victorian-era viaduct through Castlefield to the Cornbrook interchange on the edge of Pomona, at which point it travels on a new concrete viaduct through Pomona Island, over the Ship Canal, and toward Salford Quays and Media City (two major flagship developments along the Ship Canal in Salford).

Much of the tram's path travelled to – and through – property owned by Peel, who provided funding for the Metrolink extension with the condition that the tram would offer direct access to the site (Symonds Travers Morgan 1996: 10). The end result was Pomona station, along with the Cornbrook station, which occupies the edge of the site.

Reflecting concerns about Pomona and its environs' continuing existence as a wasteland, for years Cornbrook was used only for interchanges, despite the fact that it had been built as a fully serviced station. It was not until 2004, due to pressure from property developer Urban Splash, which had recently developed a number of apartments nearby, that a £250,000 'rescue plan' was devised by a working group composed of transport officials, police, city officials, and Peel instigated a number of modifications. At this point the area began appearing in the news media, cast as a problematic landscape needing swift remediation. As the local media proclaimed, 'the lure of classy city apartments will rescue a white elephant'. Proclaiming the site a wasteland, the newspaper heralded security measures – such as installing CCTV cameras and lighting to remove the 'chance for strangers to hide in the shadows' – as well as a variety of aesthetic modifications to make the site suitable for exit and entry. These included the installation of fences around scrapyards to shield their 'unsightly views', cleaning brickwork, and cutting back vegetation (*Manchester Evening News* 2004). Meanwhile, the net effect of Pomona Station had been Peel's construction of one small office building, Adamson House. To this day, Pomona is the least used station in the entire Metrolink tram network (Stuart 2013), Adamson House operates at low or no vacancy, and Cornbrook station is used almost exclusively for transfer between lines.

In 2007, Peel gained planning permission to develop 546 apartments in five waterfront buildings (varying from eight to 16 storeys in height) on 1.7 hectares, including a marina. These plans were put on hold due to the recession. As part of their £50 billion Ocean Gateway plan, which encompasses much of Liverpool's waterfront and a 58-km stretch along the Ship Canal to Manchester, Pomona is one of 50 strategic points in Peel's waterfront transformation of much of north-west England, with an ambition to 'compete against the most well-recognized and successful waterfront cities in the world, such as Vancouver and Shanghai' (The Peel Group 2009: 38). A clear illustration of the financialization of real estate (Smart and Lee 2009), Peel is largely operated from the tax shelter of the Isle of Man, with 25 per cent of the company owned by Saudi-based Olayan Group (ExUrbe 2013). Still, despite its reinsertion into global circuits of capital and grandiose aspirations, Pomona largely exists in the same condition it had since the 1970s. As Figure 12.4 depicts, it is a blank space upon which future ambitions can be projected.

Figure 12.4 The Pomona site, as photographed from the roof of Exchange Quay in Salford, 2015 (image source: Robert Watson Studio).

Pomona Island as an accidental park: wastelands as temporary commons

> The most difficult wastelands to convert are those that are also occupied and used, however ineffectively, and where the attachments, interests, and activities of the occupiers are an intimate part of the conditions of the site. Coming to terms with those conditions and allowing for the users to join the process of waste removal and rebuilding are things we have not yet learned how to do well.
>
> (Lynch 1972: 234)

With the exception of a self-enclosed scrapyard and low-occupancy office building, Pomona is a vacant plot where nothing is supposed to happen. It is a non-place of transit and passage (Augé 2009). However, like most urban wastelands on urban peripheries, the site has a multitude of informal uses. Campo (2013) described how the Brooklyn waterfront in the early 2000s became an 'accidental playground' in the interim between industrial displacement and residential redevelopment. Much in the same way, Pomona has come to exist as an accidental park. Through observing the spatial practices observed at Pomona, the signs of human activity, and discussions with individuals who value this *terrain vague*, the following section offers a sketch of the site's heterotopic (Foucault 1998; De Cauter and Dehaene 2008) character. Whether or not the users of this space have a formal right to be there, they have reappropriated it for their own uses and made a liminal site that is officially 'empty' into a temporary commons. However, all of this happens in the indeterminate interim between various stages of development. The site, which had seen industrial capitalism usurp its agrarian past, is now the subject of new forms of enclosure through the speculative restructuring dominated by finance capital.

Like other wasteland sites, landowners (Peel along with Network Rail and Transport for Greater Manchester) have made modest attempts to prevent trespassing, and as the site is largely defined by infrastructure, there are a number of barriers to entry. Despite some fencing, construction hoarding, and locked gates, pedestrian access has never proven difficult: one need only look for gaps in fences along the railway arches or broken gates. Peculiarly, one may enter the site from the Ordsall district of Salford by the Woden Street Footbridge (1873), a remnant of industrial-era connectivity, which crosses the Ship Canal and offers easy access. It is the residents of working-class Ordsall who are among the most prominent users of Pomona, many of whom consider it an extension of their neighbourhood.

Amidst the layers of transport infrastructures, with plant and animal life flourishing in the spaces between them, are the materials signs of past and present human activities. Concerns over criminal and 'anti-social behaviour' are easily illustrated through a site survey, less through the frequency of observed activities than by the prevalence of waste matter. Of course, since there is only a

Figure 12.5 A 'desire line' created by the footsteps of users entering Pomona Island from the Woden Street footbridge. The railway viaduct may be seen to the left, and the Ship Canal to the right (photograph by the author, 2010).

Figure 12.6 Passage through railway viaducts, signs of fly-tipping and proliferating flora (photograph by the author, 2010).

Figure 12.7 Railway arch shelter and entrance through slats of fence. New housing and abandoned factory in the background (photograph by the author, 2011).

minimal maintenance regime for the site, these materials accumulate over time. One matter of widespread concern is cable theft from the electrified railways, causing massive delays throughout the regional rail network. Piles of burned rubber indicate that cables are melted down at this site. On the rare occasion that the site is covered by local media outlets, it tends to be when an unidentified body is found or when a cable thief has fallen off a railway viaduct. On the site's perimeter, it is common to encounter piles of fly-tipped tires, and abandoned motorcycles and cars are not uncommon. Graffiti dominates the walls and arches of the railway viaducts: the same structures that are sandblasted and floodlit in Castlefield as monuments to industrial heritage. Accumulated litter attests to the use of railway arches for sex acts, drug use, and drinking. While lack of surveillance drives justifications for security measures and enclosure, these efforts are haphazard.

Purely in terms of visual and material artefacts, the most apparent use of Pomona is as shelter for rough sleepers, particularly within the railway arches. This is certainly not something specific to Pomona, as residual spaces of infrastructure often serve as refuge for homeless people (Rosa 2014; Tonnelat 2008). The marginality of these makeshift dwellings are juxtaposed by their proximity to the chic flats of St George's. These shelters are shielded from view by hoardings that wall off the railway arches: on the side facing the street are advertisements for upscale housing. During daytime it is much more common to encounter the signs of rough sleeping – such as sleeping bags – than to encounter homeless individuals themselves: like formal parks, Pomona's nocturnal activities differ significantly from its diurnal character.

Still, overall the activities at Pomona differ little from activities in any public open space: the difference lies in maintenance regimes, policing, and other forms of governance. Among the more common activities observed have been dog-walking, jogging, and teenagers socializing. Less frequently one can observe people fishing, bird watching, bicycling, motorcycling, gleaning wild berries, taking photographs, and recreational boating along the canals. Like formally designated public spaces, it is a place for social mixing: Pomona Island abuts some of the most deprived districts in Britain, but also some expensive new flats. Repetitive access is clearly demarcated from various entry points to the site. Conversations with recreational users suggest that the site is popularly used as a park from a broad spectrum of neighbouring residents. Elsewhere, subtle signs of repetitive movement through this space are apparent in the 'desire lines' worn into untended grass (see Figure 12.5). This path through the grass leads to what has been nicknamed 'Ordsall Beach' (Pivaro 2008a). The *Salford Star*, an independent local newspaper, ran a satirical series on the 'Costa del Salford' (a play on the Spanish Costa del Sol) along the Ship Canal, lampooning developers' desires to exploit waterfront vistas (Pivaro 2008b). Beneath the tongue-in-cheek narrative is political critique centred on Ordsall residents' desire for parkland. At Ordsall Beach,

> the naturally occurring red sands … provide the perfect platform to watch
> … blackened waters feed into the Manchester ship canal. And under the

arches [we] played and relaxed in the blistering sun, disturbed only by the 14.04 Altrincham to Piccadilly [train] hurtling above.... This place does feel like a people's beach, not fancy yet a true oasis of relaxation in the post-industrial deconstruction site.... For years we were denied what is the city's right, the river that runs through it! Now is the time to put it back where it belongs, in the hands of its citizens

(Pivaro 2008a: 32)

Threaded throughout the satirical feature was the implicit acknowledgement that temporary appropriation was a resistant but fleeting act (see Figure 12.9).

Three years later, after completing my fieldwork, groups began emerging in Manchester based on individuals' enthusiasm toward Pomona. In 2011, the Greater Manchester Ecological Unit (GMEU), which operates as an advisory service for the ten Greater Manchester district councils, produced an ecological survey of Pomona Docks. The report suggested that the site be considered for selection as a Site of Biological Importance (SBI) – a non-statutory designation that a site should be prioritized for nature conservation – based on its high level of biodiversity and cultural significance (Walsh 2011). However, as the GMEU report notes, in January of 2011 over 90 per cent of the trees and vegetation on the site were removed by the landowners, rendering the ecological survey moot. This clearance roused opposition, focused on the social and ecological signifi-cance of this supposed wasteland. Through online forums, a group of bird-watchers, architects, artists, ecologists, preservations, writers, political activists,

Figure 12.8 Man cycling along canal towpath, ducks in foreground (photograph by the author, 2011).

Figure 12.9 The *Salford Star*'s feature on 'Ordsall Beach,' shot at Pomona (image source: Pivaro, 2008).

Figure 12.10 Mature trees in 2010, prior to clearance (photograph by the author).

Figure 12.11 Pomona after clearance of plant life, 2014 (photograph by Stephen Smith, courtesy of the artist).

and urban explorers began articulating the values of the site. Local writer and curator Hayley Flynn began highlighting Pomona and its historical significance on her personal blog (Flynn 2013, 2014b), and later in the *Guardian* (Flynn 2014a), calling Pomona the city's 'alternate countryside' and a 'serene wasteland' and exalting the 'beauty of the desolate' (Flynn 2014b). Flynn, along with a growing coalition of campaigners organized a 'seed bombing' and 'protest picnic' event in March 2014 to raise awareness of the biodiversity and historical character of the site, hoping to protect the site from impending development and maintain it as a vast, wild meadow. A common perception among campaigners was that Peel's decision of clearing of the site was a 'scorched earth policy' driven by concerns that for Pomona could be listed as an SBI, which could limit its development potential (Keeling 2014). However, since the Trafford and Manchester councils would ultimately have the authority to determine this status and Peel has been unwilling to comment, it is unclear whether this suspicion is warranted.

Also in 2014, filmmaker George Haydock directed a short documentary film entitled *Pomona Island*, inspired by his fascination that 'through the cracks of hyper-development and regeneration, here lies a totally unmanaged, largely un-used and unnoticed area of land.... It just exists in a strange limbo between its former industrial use, and its inevitable destruction and future development' (interviewed in Flynn 2014b).

The current fascination with urban wastelands, ruins, and derelict urban spaces also adds an additional level of complexity to debates around the wastefulness and value embedded in Pomona. While Peel is cautious to acknowledge that there is any existing recreational or ecological value to the site as it currently exists, it is actually able to commodify the area's current decay to location scouts looking to shoot films or television shows. Through its website (filmandtvlocations.co.uk), The Peel Group leases the site to film crews, offering them exclusive access to an authentic post-industrial wasteland. In this sense, even when the site has no formal use, it still generates value for the landholdings company precisely because of its status as a wasteland.

The campaign to 'save' Pomona, or at the very least appreciate it as a precarious, temporal landscape, has also been savvy in harnessing the steadily increasing interest in Britain's urban wastelands, and in particular, narratives about biodiversity and the need for publicly accessible green space in Manchester. At this point, it remains unclear whether these activists will sustain their efforts to challenge the inevitability of the site's redevelopment, or whether these activities will simply celebrate the site's uniqueness in the face of its impending erasure. What is certain is that a central claim of these artists, urban explorers, and ecologists is that the site has significant value in its current state.

Use values, exchange values, and aesthetic values: what Pomona tells us about wastelands

The 'waste' in wastelands contains a dual meaning: an evaluation of the physical condition of a particular site coupled with the economic value it generates. The economically un(der)productive and informal uses of these spaces, often an expression of commoning (Caffentzis and Federici 2014), are often downplayed or ignored completely. The case of Pomona illustrates a number of points about the ambiguous and multiple meanings of urban wastelands.

Urban wastelands often tend to be defined as such when they are perceived to be serving as a barrier to property-led redevelopment: the outward expansion of a city centre, the 'reclamation' of formerly industrial urban waterfronts for luxury housing, and more generally, gentrification. In other words, a site becomes a wasteland within official discourses as a justification for its reconfiguration, much as the term 'slum' has often served to describe an urban neighbourhood slated for residential displacement and urban renewal (Gilbert 2007).

Di Palma concludes her cultural history of wastelands with a cautiously hopeful outlook toward the future of post-industrial leftovers:

> Wasteland bears witness to [our] actions; it is our conscience, our terrain of contestation. As a space of resistance, of challenge, and, ultimately, of possibility of change, wasteland has the potential to be the landscape paradigm for our uncertain and troubling times.
>
> (Di Palma 2014: 244)

With the wild popularity of New York's High Line (Lindner and Rosa forthcoming) – a linear park atop a disused railway viaduct – along with the new Tempelhofer Freiheit and Park am Gleisdreieck in Berlin, the aesthetic appeal and design possibilities for such post-industrial sites are evident. Clearly, understanding urban wasting is a process (Lynch 1990; Southworth 2001) is an essential step forward in understanding the spatio-temporal and ecological elements of urban wastelands. Still, there is a danger in (a) naturalizing the mobility of capital and labour that lead to disinvestment and obsolescence and (b) recognizing that the superficial celebration of wastelands and their affective qualities can be used as a tool for further property speculation (Loughran 2014). Furthermore, as much as architects and landscape designers increasingly celebrate the ecological, aesthetic, and affective values of urban wastelands, the case of Pomona illustrates the lack of agency that designers have. Even if they continue to successfully promote a wasteland aesthetic as a new trend in landscape design, it is a consensual partnership between landholders and local councils that will determine the future of the site.

At this point, the most radical idea for Pomona would be to leave it in its current state. Ironically, this has been the most consistent treatment of the site since the closure of the docks, yet its continuation as an open space has been completely foreclosed as a viable future. However, under the logic of neoliberal planning, even if public support is behind maintaining an urban wasteland as a

communal space of natural conservation and recreational usage, it must be justified through commodification. The most common economic justification for such
conservation is the encouragement of tourism, itself assigning economic value
(Hall 2013). In this particular case, a small group of enthusiasts – varying greatly
in interests but sharing an appreciation for Pomona – have begun to articulate an
alternative vision for the site, based on values of ecological distinction, aesthetic
richness, heritage, and the desire for a less manicured and regulated type of
public space. 'Its wildness', according to a Pirate Party UK member and
defender of Pomona, 'is for many a blessed contrast to the deathly dull of places
like Spinningfield [a recent public–private development in Manchester]' (Kaye
2014). However, the unabashedly entrepreneurial city council of Manchester is
vested in the transformation of the site into housing, despite the fact that this
ambition conflicts with its alleged commitment to green urbanism. The city itself
casts the area directly next to it, Cornbrook, as an 'intimidating' urban wasteland, disseminated through press releases ventriloquized by the local media
(Williams 2014).

Pomona also emphasizes the fact that a site's dereliction or apparent neglect
may belie, or even signal, significant economic interest. One needs look no
further than the string of aborted masterplans to see that Pomona is the subject of
considerable interest for large-scale redevelopment. The actions of Peel – particularly the intentional destruction of Pomona's well-established and biodiverse
wildscapes – offer a unique dynamic. Peel's stated concerns that continued non-
intervention into the landscape would be 'unfair' to wildlife (personal interview
with anonymous informant 2014), or that allowing wildlife to flourish could lead
to a legal claim of its ecological significance, led to a strategy of repeatedly
uprooting the non-human life that has come to occupy the space. Beyond taking
a stance of maintaining Pomona as an wasted and 'wild' urban void, Peel actively participates in its degradation.

Even as Pomona has a long history as a pleasure garden and was earmarked
for the establishment of a public park in the 1980s and 1990s, all alternatives to
dense, market-rate housing and commercial development on the site have been
foreclosed without any sanctioned public dialogue. In Manchester, political
antagonism and democratic participation in spatial governance have been
replaced by a neoliberal, post-political planning process focused on elite consensus (Allmendinger and Haughton 2012). With the majority of planning
schemes occurring in Manchester occurring as a partnership between the city
council and property developers, it is against the city's interest to entertain any
alternative visions for the allocation and use of land in the city. Therefore, any
real hope of contesting the plans to turn Pomona Island into an exclusive residential and commercial enclave could only occur through highly visible and
vocal contestation.

There is an ambivalent politics of visibility within spatial practices that seek
to emphasize the desirable, heterotopic qualities (De Cauter and Dehaene 2008)
of urban wastelands. This is because, by the very act of making these sites
visible, they could easily sow the seeds of destruction for any qualities

appreciated in that site: relative freedom, lack of regulation, or biodiversity. It is curious to wonder whether groups emphasizing the site's ecological and recreational values might simply be fuelling the 'rediscovery' of Pomona if they are unwilling to engage in a protracted and committed battle against the site's redevelopment. As Haydock's documentary emphasizes, the island's enthusiasts

Figure 12.12 An announcement for a tour of Pomona led by Hayley Flynn, as posted on her blog *Skyliner* in 2013 (image source: Hayley Flynn).

have varying visions of what the site should be, and whether it should simply be appreciated in its temporary state or fought to conserve. The tactics of groups celebrating and appropriating Pomona as a post-industrial wasteland do not suggest that there is a concerted and organized effort to contest the eventual enclosure of Pomona. According to Hayley Flynn,

> my aim is to encourage the public to use it whilst they can, with a view to making developers consider the worth of wild green space, but I don't hold any hope that Peel would even consider that as a use. I just want it to be utilized by residents and dispel any fears people have of the area.
>
> (Personal interview 2015)

She says that it is 'begrudgingly' accepted that any public use will be temporary.

While it is unclear whether the actions of the ecologists, public space advocates, artists, and designers will have any effect on the development of Pomona Island, they demonstrate the fact that the site's value is precisely because it is *not* improved. If anything, the repetitive clear-cutting of the site produces a result that is, by many standards, a blight rather than an improvement. However, to Peel Holdings and to the municipal authorities of Trafford and Manchester, this is still an unwelcome criticism of the way that redevelopment occurs in the metropolitan area. Various forms of resistance call into question the assumption that the site is empty of all appreciation or use, or that it has value beyond its position as a staging ground for future construction. Treating a site as a blank space becomes complicated when wastelands themselves become appreciated and considered to be worthy of protection from future development. I anticipate that these tensions will continue to emerge as an important element of future political contestation around the enclosure of urban wastelands that, of course, were never valueless. This can only happen in the moments when the values of urban spaces are articulated outside and beyond the logic of capital accumulation. However, this will require a shift from the celebration and romanticization of (temporary) urban wastelands to active contestation around the presumed consensus of market-oriented urban planning.

13 On Beckton Alp

Iain Sinclair, garbage, and 'obscenery'

Niall Martin

From the summit of the last remaining slag heap of the former Beckton Gasworks in East London, writer and filmmaker Iain Sinclair surveys the scene.
Looking west, back to the City, the financial centre of millennial London, he
describes something like an epiphany:

> Leaning on a creosoted railing London makes sense. There is a pattern, a
> working design. And there's a word for it too: Obscenery. Blight. Stuttering
> movement. The distant river. The time membrane dissolves, in such a way
> that the viewer becomes the thing he is looking at.
>
> (Sinclair and Matthews 2002: 22; Sinclair 2004: 190)

We can make sense of Sinclair's making sense in a number of ways. From
Beckton, the site of a major sewage treatment plant and of the former gasworks
that between 1879 and 1969 met the energy needs of a rapidly expanding city,
the complexity of London resolves itself into the simple dynamics of an organism. The apparent chaos of the metropolis reveals itself to be governed by ele-
mentary flows of consumption and excretion. The city needs energy and
produces garbage: breathe in, breathe out. This vision of the city as organism is
repeated elsewhere in Sinclair's work when he writes, for example, of the city
which 'snorts human meat through metalled tubes. And later exhales the de-
energized husks, its wage slaves' (2005: 133). Garbage, as waste or refuse, in
other words, invokes powerful metaphors of organic order.

So, too, at Beckton Alp, Sinclair stands on a spoils heap composed of coal
shipped from mines in the north of England for the manufacture of chemical
products, that were then sold throughout the colonial market (Townsend 2003).
He stands on the remnant of a nineteenth-century industry which was fully integrated into a colonial system of procurement, manufacture, and distribution. A
system whose disappearance consequently signals London's own altered position
in the world: its transformation from imperial metropole into a node in Saskia
Sassen's network of global cities (1991). As such, this East London slag heap
makes sense of London within the shifting geographies of the global economy.

Equally, as a monument to an obsolete coal-based manufacturing industry,
Beckton serves as a precise historical marker of the emergence of the petro-dollar

Figure 13.1 From Beckton Alp looking north-west (photograph by the author).

as a new force within global economics in the 1970s, a force which continues to shape London's features both as a financial centre and property market to the present. Looking back from Beckton to the millennial City, Sinclair thus looks back from the past towards the future, from the city of matter to the city of signs.

Viewed from the perspective of this slag heap, then, it is possible to discern a 'working design' that makes sense of the city in terms that are at once systemic, economic, and historical. Here, however, I want to try and make sense of Sinclair's making sense by concentrating on the trope underlying this epiphany: the familiar trope that waste, refuse, and garbage in general provide a privileged perspective from which to make sense of a world that would otherwise appear fragmentary, contingent, without design. Just why is garbage, in aesthetic terms, so peculiarly valuable?

It is a trope which owes its familiarity to the fact that it seems to be written into the DNA of European modernism. From Walter Benjamin's 'rubble of history' (via Paul Klee's 'Angelus Novus') to T.S. Eliot's *The Waste Land* and Marcel Duchamp's ready-mades, garbage constitutes one of the key conceptual resources deployed by artists and theorists of modernity to make sense of their social and historical reality. Indeed its cultural reach is registered in the demotic wit that transforms an East End slag heap into an Alp – a mock-heroic manoeuvre which relies on Romanticism's reclamation of rocky wastelands as a proper terrain of the aesthetic some centuries earlier (Hope-Nicholson 1959). It is a trope, in other words, which by virtue of its very ubiquity can be recognized as constitutive of the Romantic/Modernist tradition in which Sinclair stands and upon which he reflects.

To investigate the reasons for and implications of this ubiquity I want to distinguish between two narratives of waste, one which figures waste as relational and another which focuses on its persistence or revenance. By considering their combination in Sinclair's Beckton epiphany I want to indicate some of the ways in which these narratives open up and foreclose different configurations of global space.

Waste relations

The first narrative I am going to term the narrative of waste relations. In this narrative, waste, like Mary Douglas's dirt, is always relative. Just as dirt is simply 'matter out of place' (Douglas 2002 [1966]), so too waste lacks any intrinsic, material dimension and becomes waste only in relation to a use or exchange value, calculated in relation to a particular subject. Waste is always waste from the perspective of someone, or something, or some construction of value. In this narrative, waste typically takes on a deconstructive function insofar as it serves to expose the interdependence of the symbolic categories through which waste and value are constructed.

Among the many commentaries on this deconstructive story of waste, Barbara Babcock's account of 'symbolic inversion' remains among the most potent. Her observation that 'what is socially peripheral is often symbolically central' (1978: 32) describes the central dynamic of this narrative and alerts us to the many ways in which the symbolic role of marginalized categories, such as garbage, assumes an importance out of all proportion to their imputed social worth. The Sex Pistols' transformation of dustbin bags into items of haut couture (McClear 2013) provides a familiar example. John Lydon's demand that people 'wear the garbage bags' (Temple 2000) as a response to the escapism of youth fashion in a world of social deprivation created, of course, one of the icons of tourist-London. It is the same formula that inspires Peter Stallybrass and Allon White's (1986) study of the central role of garbage in the reformist discourse of nineteenth-century British urbanism. It explains why for Mayhew, Chadwick, Engels and even Marx, '[t]he melodramatic coercion of extreme opposites into close intimacy ... becomes the ultimate truth of the social' (Babcock 1978: 141), with the result that the sewer becomes the conscience of the city.

However, if we are to understand why it is that 'Mayhew fixates upon bone-grubbers, rag gatherers, 'pure'-finders (collectors of dog shit), sewer-hunters, mud-larks, dustmen, scavengers, crossing-sweepers, rat killers, prostitutes, thieves, swindlers, beggars and cheats' (Stallybrass and White 2007: 269), perhaps Spelman provides a simpler answer when she writes that insofar as we define ourselves negatively through what we abject 'trash provides an epistemologically privileged resource for understanding what individuals and communities are all about' (2011: 323–4). Or, as Myra Hird puts it, 'what remains after our disgorgement is what (we want to) consider our real self.... In other words, we know ourselves through waste' (2012: 456). Waste, it seems, produces ontology.

But more specifically, insofar as concepts of waste in this narrative are always bound up with concepts of value, it follows that the more sophisticated the theory of value, the greater the potential for the forms of waste. Marx's analysis of capital and the commodity form, for example, presents us with a variety of distinct forms of waste. Most obviously there is waste related to use value – the spectacular waste of the empty mansions in London's Bishops Avenue or 'billionaires' row' to take a recent example (Booth 2014). However, in terms of exchange value, waste may take the form of a commodity's failure to realize its potential for profit – a low-rent building on a prime location, or equally, in a discrepancy between the invested value and depreciating exchange value of a commodity – the 'blight' signalled in Sinclair's survey of millennial London.

Within Marx's analysis of capital's production of value, however, waste takes on a function which seems generative or originary to a degree that is almost metaphysical. Thus in his account of the historical emergence of capitalism through the commodification of the surplus labour of landless populations, waste is placed at the origin of capitalist value insofar as it is labour which is wasted from the perspective of the labourer, producing the surplus value that allows the accumulation of capital (Yates 2011: 1688). So too, in his historical account, it is the production of human populations as waste to be cleared from land which could be more profitably devoted to sheep farming that initially primes the pump of European capitalism (Marx 1990: 877–95). The position of waste at the historical origin of the narrative of capital's development is complemented in his analysis of the logic of the commodity form. Here waste has a similarly generative function insofar as the commodity only comes into being at the moment of consumption. So an unsold item 'goes to waste' in the sense that the labour and investment it embodies never realize their exchange value, but so too,

> [a] product becomes a real product only by being consumed. For example, a garment becomes a real garment only in the act of being worn; a house where no one lives is in fact not a real house; thus the product unlike a mere natural object, proves itself to be, *becomes*, a product only through consumption. Only by decomposing does consumption give the product the finishing touch.
>
> (Marx 1993: 91, original emphasis)

From a Marxist perspective Beckton usefully condenses several varieties of waste into one location. As a slag heap, the Alp was produced by an industry dedicated to extracting use value from waste: to producing new commodities (including the creosote on Sinclair's railings) from the waste products of other industrial processes, primarily the production of gas from coal (Townsend 2003). As the waste of an industry born out of the transformation of waste, the Alp thus represents a limit to the power of human ingenuity, its ability to discover use value within matter. It is in this material sense, a form of absolute waste: the practical limit of a technocratic faith in better living through chemistry. However, the obsolescence of this industry itself is consequent not on a loss of use value but on its loss of exchange value in the face of competition from cheaper petroleum-based technologies and the sudden availability of North Sea gas. As such, Sinclair's slag heap embodies waste both as use and as exchange value: it represents a waste degree zero.

This sense of the Alp as a monument to absolute waste is reinforced by the failure of subsequent attempts to make it pay by reincorporating it within the urban economy as real estate. At the time of Sinclair's first visit, the Alp had been briefly reinvented as an artificial ski-slope, but two years later it had shrugged off this incarnation: 'The ski-slope had been dismantled, asset stripped.... The scam was discontinued' (2004: 186). In February 2014 it remained an unreclaimed space, nominally fenced off for the protection of the public, but in fact used for a variety of purposes including mountain-biking, dog-walking, partying, letting off fireworks and providing ironic captions for the London landscape.

However, as the waste relations narrative would suggest, and as Sinclair's own text demonstrates, even as waste degree zero, Beckton continues to produce value. Within Sinclair's text, Beckton rises again as a symbol of obsolescence, and as a monument to the obsolescence of the city that produced it. And, of

Figure 13.2 'God Save the Scene' (photograph by Paul Caplan, courtesy of the artist).

course, Sinclair is not alone in exploring the semiotic and aesthetic possibilities of dereliction. In this case, the end of Beckton as a site of industrial production raises the curtain on a particularly vigorous afterlife in the visual culture of the late twentieth century. A partial credit list reads as follows:

1975 Feature film, *Brannigan* (Dir. Douglas Hickox)
1981 Feature film, *For Your Eyes Only* (Dir. John Glen)
1984 Feature film, *1984* (Dir. Michael Radford)
1985 TV movie, *Max Headroom: Twenty Minutes into the Future* (Dir. Annabel Jankel and Rocky Morton)
1986 Feature film, *Biggles: Adventures in Time* (Dir. John Hough)
1986 Film/music video, The Smiths, 'The Queen Is Dead' (Dir. Richard Heslop/ Derek Jarman)
1987 Feature film *Full Metal Jacket* (Dir. Stanley Kubrick)
1987 Music video, The Outfield, 'Since You've Been Gone'
1988 Film *The Last of England* (Dir. Derek Jarman)
1990 Music video, Loop, 'Arc-Lite'
1994 Music video, Marcella Detroit, 'I Believe'
1997 Music video, Oasis, 'D'You Know What I Mean?' (Dir. Dom&Nic)

Its different functions as locale will be considered below; here I want to note simply that in this phoenix-like transition from the material to the semiotic, or industrial to the cultural, we can see a crude rehearsal of one of the principal themes of twentieth-century accounts of the 'aesthetic': the idea that the aesthetic is itself constituted out of an always volatile relation between competing notions of value and waste. That volatility is of course grounded in Kant's identification of the aesthetic with the principle of disinterest: his insistence that art is art because uniquely in a world dominated by value for another, it possesses the quality of being only for itself. In the twentieth century the forms of interest in this theory of disinterest are scrutinized by, among others, Theodor Adorno and Pierre Bourdieu. For Adorno, it is art's apparent exemption from the systems of instrumental and exchange values that govern the rest of social life that simultaneously underwrites its promise of an (illusory) realm of freedom, and fuel the outrage of puritans 'of all stripes' (Adorno 1999: 122), who can see in this refusal only an indication that the meaning of art is indeed that it is worthless. So, too, for Bourdieu, art cannot be thought without the concept of waste. Produced within a 'restricted economy' governed by the principle 'loser wins', aesthetic production is structured around the idea that art is defined by (a) its lack of utility or use value and (b) the artist's refusal of exchange value in the form of commercial success: 'the fundamental law of this specific universe, that of disinterestedness ... is the inverse of the law of economic exchange' (Bourdieu 1993: 164).

As Robert Bond points out, Sinclair's text investigates this uneasy correlation of waste and the aesthetic in a number of ways. It is evident, for example, in his concern with the 'reforgotten' (2005: 30): in poets such as Anna Mendelssohn and Harry Fainlight whose wilful 'obscurantism', or uncompromising dedication to their own 'voices' constitutes a degree of artistic integrity and purity that ensures they will never attract mainstream attention, and who as such confront Sinclair with his own 'failing to fail' (Sinclair 2011: 147). In this commemoration of the wasted lives of artists who are necessarily 'reforgotten' as a consequence of their purity, his own text becomes a form of waste, an unrequired supplement.

The sources of this uneasy relationship between value and garbage can of course be located in the Kantian conception of aesthetic disinterest and art as that which is without instrumental or exchange value. Sinclair's engagement with this Kantian formulation of the aesthetic is complicated by his acute sense of the dynamic and often symbiotic relation between art and capital, garbage and value. This is evident, for example, in the Beckton epiphany itself, which, as text, occurs first in *White Goods*, in 2002, where it describes a direct perception or revelation of meaning: an epiphany. However, it is then recycled as a quotation attributed to another author in *Dining on Stones*, in 2004. In this degradation from direct perception to the kind of fragment of language that Kirsten Seale terms 'textual refuse' (Seale 2005) it mirrors Marx's metaphysical account of the relation between commodity and waste. Textualization consumes the event of perception. As perception becomes text, it is inducted into a system of circulation where it inevitably succumbs to the dual logic of quotation, which simultaneously preserves and consumes the quoted as a fragment that points to its own fragmented character. In this, it reveals the doubled character of the modernist text as itself a site of both production and consumption, presenting itself as both work and midden: an assemblage of fragments shored against its own ruin.

It is in his accounts of Jock McFadyen, however, that Sinclair provides his most direct reflections on the relations between the aesthetic, garbage, and capital. Caricatured as Jimmy Seed in *Dining on Stones*, Sinclair presents McFadyen as an artist who navigates the opportunities thrown up by the troubled border between art and capital with a brutally pragmatic efficiency. He describes his *modus operandi* thus:

> Sell a painting, buy a share in a slum. Sell the renovated flat and option a burnt out pub. Photograph the pub, copy the photo. Sell the painting to a hustler who risks a wine bar on the same site. Sell heritage as something to hang on the Gents.
>
> (Sinclair 2004: 10)

As an artist Seed/McFadyen paints the world that will subsequently appear in the developer's property portfolio; he performs the preliminary imaginative investment necessary to prepare, or territorialize, the world for investment capital:

> I imagined when McFadyen began to blaze away with his camera, that he was gathering material for a future painting. Nothing of the sort. The artist was sussing a property. These days the huge canvas is of secondary interest.

> Paintings are no more than blow-ups of estate agents' window displays. They're done, the best of them, with a lust for possession. Speculations that got away.
>
> (Sinclair 2003: 48)

In this caricature Sinclair explores another congruence between the operations of capital and the operations of art concealed within the romantic economy of the aesthetic. In collapsing the distinction between imaginative and commercial speculation, McFadyen effectively presents us with that general spirit of 'collusion between capitalism and the avant garde' that Jean Francois Lyotard suggests serves as a convenient marker of the postmodern (1991: 105).

Cast as an abortive attempt to find words to accompany McFadyen's postmodernist painting, which inevitably leads to a reflection on the continuing devaluation of the word in an increasingly visual culture, the Beckton narrative soon turns into a meditation on the failure of representation in general, and leads ultimately to the conclusion that '[t]he artist is redundant. As is the reporter. The photograph. The memory prompt. Useless. We're still on the inside of the outside, searching for cracks. Trapped in an envelope of tired flesh. View is raw and absolute and unappeased' (Sinclair and Matthews 2002: 22). In this perception of redundancy, the waste-as-relation narrative reaches its logical terminus. No longer able to act as the conveyer of cultural capital or as arbiter between the memorable and the ephemeral, nothing is left to the artist but to contemplate his own redundancy. To figure his own work as a cultural midden that gathers together scraps of culture in its own distinctive landfill otherwise known as the classic modernist text.

Within the narrative of waste relations, in other words, Beckton represents the general ambivalence of the aesthetic as a category uneasily caught in the suspicion that that which is without value for another is simply without value, that art for art's sake is simply another term for a load of old rubbish. However, it is out of the crisis of this narrative in which the artist confronts her own redundancy, that my second provisional narrative of garbage emerges, the narrative of garbage as that which cannot be disposed of, of garbage as the revenant. In the following section I want to consider the ways in which this narrative produces and reflects upon the spaces of globalization.

Refuse as refusal

Where the first narrative of waste as relation finds its motto in Babcock's assertion of the symbolic centrality of the socially peripheral, the second can be read as an elaboration of Jean Baudrillard's adage that 'the end of history is, alas, also the end of the dustbins of history' (1994b: 26), for, in this narrative, garbage figures primarily as a problem of storage. No longer that which is simply out of place, garbage becomes that which confronts us with its inability to be placed: with its irreducibility and its indeterminacy. Rather than being determined in relation to a particular subject position, this is the story of a waste which, in insisting on its inability to be disposed of, constantly undermines the possibility of asserting stable subject positions. It is a narrative which seems particularly attuned to ideas of globalization organized around the assumption that the global extension of a single system of commodity exchange signals the 'end of history' (Fukuyama 1989) and that we now inhabit a commonality which is inescapably interconnected and inescapably post-human.

The potential manifestations of refuse as refusal to be placed are multifarious. It is refuse as refusal, for example, that Myra Hird invokes in her account of the landfill planners' struggle to contain the volatile chemical mix known as leachate. For, as leachate, Hird points out, garbage 'moves into and through plants, trees, animals, fungi, insects and the atmosphere'. Assembling 'geo-bio networks ... it travels vertically and horizontally within landfills, and continues to travel when it leaks beyond landfill cells' (2012: 457). As leachate garbage becomes radically indeterminate. In its 'exuberance' it dissolves any distinction between what is abjected as waste and what is constitutive as environment.

As that which destroys the possibility of any distinction between subject and object, self and environment, refuse as refusal presents itself first and foremost as a form of radical indeterminacy, as a refusal to be known. Just as waste breaks out of its landfill cells, Hird argues, it leaks from the conceptual categories constructed for its definition and containment:

> Just as engineers endeavour to determine landfill containment, so too do societies attempt to effect forms of ideological, symbolic and social containment by rendering waste determinate. Waste is materially contained through human disposal practices of household waste sorting, curbside pick-up, recycling, landfilling, incineration and the like. It is ideologically, symbolically and culturally contained through these material practices, as well as legislation, surveillance, public education, health discourse, nation-building rhetoric and so on.
>
> (2012: 465)

Clearly any engagement with refuse as refusal, as that which in its exuberance dissolves the boundary between the human and the inhuman, has implications for thinking the relationship between garbage and the global. Timothy Morton, for example, draws one important conclusion when he writes that

> [h]uman society used to define itself by excluding dirt and pollution. We cannot now endorse this exclusion, nor can we believe in the world it produces. This is literally about realizing where your waste goes. Excluding pollution is part of performing Nature as pristine, wild, immediate, and pure. To have subjects and objects, one must have abjects to vomit or excrete (Kristeva).
>
> (2010: 274)

Refuse as refusal to be placed refuses to work for the production of subjects and objects, and instead becomes something more like a threshold or a zone of mediation and transformation. In the remainder of this chapter I want to explore the ways in which this conceptual leakage works to interrupt and disturb formations of the global arising from that already volatile relationship between garbage and the aesthetic explored in the waste relations narrative.

Ruin porn and the dustbins of history

As a product of global space, Beckton appears most obviously as an example of the processes of deindustrialization associated with the globalization of production from the 1970s. In this context it becomes a metonym for the widespread destruction of Western manufacturing industries, a portent of the ruins of Detroit and of countless other manufacturing towns that constitute the global rust belt. Inevitably then, in drawing our attention to the aesthetic potential of the Alp as a privileged viewpoint on millennial London, Sinclair is also drawing our attention to the uneasy cultural associations conjured up in the contemplation of the 'deindustrial sublime'. Associations which are captured in the more pejorative description, 'ruin porn' and insinuated in his own neologism, 'obscenery'.

Just what it is that constitutes the obscenity of ruin porn is neatly summarized by Tim Strangleman when he notes that criticism of the aestheticization of the material remnants of deindustrialization takes two main forms. Ruin porn is attacked, (a) for promoting a sentimental view of the past 'which is in danger of crowding out more critical attempts to make sense of the deindustrialization process' (Strangleman 2013: 23), and (b) as a form of voyeurism in which middle-class 'explorers' armed with 'telephoto lenses, French theory and poetic notions' (Clemens 2011: 253) replicate the Grand Tours of the eighteenth century and in so doing '"other" the strange industrial world they visit' (Strangleman 2013: 24; High and Lewis 2011: 43).

Running through both the critiques of 'ruin porn' outlined by Strangleman is a sense of garbage as defined by its relation to the proper: as the worthless that serves to produce the valuable and the spaces that are associated with being valuable. Rather than a site of waste, the industrial ruin is read as a site of loss, or memory that produces its own waste, in this case in the form of the 'coffee-table book'. Displaced from the scene of economic waste to the aesthetic, garbage still serves its traditional function of marking the distinction between the valued and the abject. On the one hand there is a properly 'critical' relation to history, on the other the coffee-table book produced for the consumption of passive or voyeuristic 'spelunkers' or 'middle class' browsers. The notion of ruin porn, which arises from reducing Beckton to a signifier of deindustrialization, in other words, preserves the role of garbage as the producer of fixed subjects and subject positions (middle class, voyeur, inauthentic, outside versus working class, labourer, authentic, inside) in locatable places (workplace, place of consumption, factory, coffee table). In this narrative, the ruin as waste organizes a comfortable moral tableau, separating true art from commodity, the critic from voyeur or consumer. As such, it is an understanding of ruin porn which is firmly rooted in the waste relations narrative of garbage.

In reading Beckton simply as a sign of deindustrialization, in other words, we foreclose other possible narratives of globalization that emerge when we consider garbage instead in its guise as the ill-disposed. In Sinclair's account we can recognize at least two ways in which the Alp as a place which cannot be placed works to present alternative perspectives on the dispositions of global space. First, it interrupts through its obduracy. As the remainder of the remainder, the embodiment of that which refuses to be assimilated into an instrumental economy, the Alp stands out in a landscape of 'retail landfill' (2003: 4) characterized by its insubstantiality. In *White Goods*, for example, Sinclair travels to the Alp across a terrain whose specificity is being absorbed into the homogeneity of globalized imagery. He approaches the Alp on a journey through a city which seems to be literally disappearing:

> Out here, the missing slabs of the map didn't matter. Everything was missing. Everything was transitional, a series of flats, hoardings that transformed luxury goods – by a system of revolving panels – into obtrusively-breasted young women. Real buildings, ex-industrial, were less convincing than the replicas: deserts, oceans, gas stations in New Mexico. A post-nuclear clarity. Fault lines radiating out from monster boards onto the camber of the road.
>
> (Sinclair and Matthews 2002: 13)

And although Sinclair's visits predate the crisis of 2008, the conditions of that crisis are retrospectively readable in this depiction of a landscape overdetermined by images of an American elsewhere that will turn out to be more real than anything within the substantial world offered by London. In this terrain where 'real buildings are less convincing than their replicas' the Alp stands out by virtue of its refusal to be assimilated into an instrumental economy. In its obduracy, its refusal to be reincorporated, the Alp reveals the correlation between mobility and value within an economy predicated on the circulation of information. Within this economy, the immobile and the untranslatable or non-fungible becomes an excess which constitutes itself as another model of absolute garbage – a garbage which cannot be contained, not because of its mobility, but by virtue of its excessive immobility.

Beckton's apparent immobility, however, is not the only source of its power to disrupt the production of global space. Its potency in this respect is also suggested when Sinclair turns his attention to Stanley Kubrick's reinvention of Beckton as Huế City for the battle scenes of his 1986 film *Full Metal Jacket*. In choosing to film the battle for Huế City at Beckton, Sinclair suggests that Kubrick effectively establishes an equivalence between the two cities, and thus lays waste to the specificity of both Vietnamese and English culture. In so doing he demonstrates that, from the perspective of the imperialist, all colonies are interchangeable, and that within the new global configuration the colonized is indiscriminately and universally 'the local', or more accurately, the local's claim to specificity: 'Call Beckton Vietnam if you want, but you can't summon a vanished culture with a truckload of palm trees' (Sinclair 2004: 187). Kubrick's equation of London with Huế is rendered all the more ironic by the fact that Sinclair views *Full Metal Jacket* as another episode in Hollywood's protracted rehearsal of America's Vietnam trauma (2004: 187), while the ability to eliminate the distinction between places, and to assert the interchangeability of colonies, testifies to Hollywood's effective victory in the battle for global dominion at the level of spectacle. Victory in the mediated landscape of the late twentieth century, as Baudrillard (1995) suggested, consists precisely in the ability to superimpose your trauma upon that of others – whether the inhabitants of Huế or those of Beckton who remembered the German bombing raids of the Blitz.

In drawing our attention to the act of erasure concealed within Kubrick's denunciation of war and the war machine, Sinclair also alerts us to the ways in which becoming waste entails the loss of specificity and of identity. Kubrick's film exploits the anonymity of Beckton as a site of dereliction because, as a site of dereliction, it has become detached from any recognizable locale. So too garbage belongs to the global to the degree that becoming garbage entails the surrender of difference and individual identity. From this perspective, garbage becomes the threshold of in/visibility: as that which cannot be disposed of it becomes a medium for stories that could not otherwise be told. In the case of Beckton Alp this is particularly marked for the derelict gasworks that is simultaneously an icon, a familiar presence on the screens of the late twentieth century, and at the same time an icon which is without location, which is unrecognized, unlike for example, the derelict Battersea power station or the converted Hoover factory on the A40.

In this, Beckton, as the ruin produced by globalization, alerts us to the presence of an alternative, subliminal, London whose story can be written only in the space occupied by the disregarded. In his identification of Beckton as the marker of an unacknowledged London in Kubrick's rehearsal of American colonial trauma, Sinclair alerts us to its subliminal presence within the visual culture of the late twentieth century. Tracing the plant's distinctive silhouette across the small and large screens of the 1980s and 1990s reveals that Beckton, while unrecognized and unremembered, serves as a site for the production of a strikingly complex narrative of nation and cultural memory.

Reading through the filmography noted above shows that after debuting as a setting for the action movies *Brannagan* (1975) and *For Your Eyes Only* (1981), starring John Wayne and Roger Moore respectively, it quickly takes on dystopian associations in Michael Radford's *1984* (1984) and Annabel Jankel and Rocky Morton's made-for-TV, cyber-punk film, *Max Headroom: Twenty Minutes into the Future* (1985). Emerging out of the white noise of the opening credits of *Max Headroom*, it also becomes identified with a grunge/punk aesthetic which is repeated in pop videos throughout the 1980s and 1990s. No longer a place, Beckton becomes a playground for video 'special effects' and image manipulation.

The temporal displacements involved in this transformation of an industrial ruin into the favoured locale of a dystopian imaginary are then further amplified by Richard Heslop's manipulation of film speeds in the opening section of a 13-minute film for The Smiths, 'The Queen Is Dead' (1986).[1] Beginning with black-and-white archive footage invoking a nostalgic England of council estates and jubilee parties, the film shows jerky, heavily processed images of Beckton, while an androgynous figure in a 1940s-style boy's outfit climbs over a perimeter wall to graffiti the message 'The Queen Is Dead' before reappearing, later in the video, in a blue silk dress.

Kubrick's transformation of the gas works into Huế City in 1986 thus intervenes in an already complex site of cultural reflection on the relationship of the national and the global and the present and the past. It is, however, an intervention which adds another layer of material and cultural memory to the site, with elements of the Huế City decor surviving into Jarman's *Last of England* and Loop's video for 'Arc-Lite', while, in the big-budget helicopter choreography of Oasis's 'D'You Know What I Mean?' (1997) video, directors Dom&Nic fuse the by now familiar post-apocalyptic association of the site with nostalgic references to Kubrick's *Full Metal Jacket*, in particular its final scene, the 'Mickey Mouse march', where Kubrick's Marines become Oasis's festival-goers.

As a site for mapping contested gender identities onto a queered version of national history, a site that melts the 'time membrane' separating dystopian and nostalgic imaginaries, and morphs into a site of nostalgia for its own invisible history, Beckton's engagement with the global cannot be reduced to the simple narrative of deindustrialization. Rather than a straightforward case of nostalgia for an imperial city or the class certainties of a vanished industrial economy, Beckton, with its fusion of the dystopian and the nostalgic, emerges as a site for the expression of a fundamental ambivalence, both toward the past and toward the present. It figures a city whose passing, while lamented, cannot be regretted – a relationship towards both past and present which will not resolve itself into any easily formulated attitude or perspective.

It is in this refusal to be resolved that we can locate Beckton's status not as ruin porn but as 'obscenery', for the obscene as the revelation of that which should have been kept hidden denotes precisely the collapse of the proper distance between subject and object that produces perspective and the possibility of scenery. Again Baudrillard is a useful guide in this respect:

> Scene and obscene do not of course have the same etymology, but it is tempting to connect the two. For as soon as there is a scene or a stage, there is gaze and distance, play and otherness. The spectacle is bound up with the scene. On the other hand when we are in obscenity, there is no longer any scene or stage, any play, and the distance of the gaze is abolished.
>
> (2003: 27)

Rather than the obscenity of turning a place of industry into a site of aestheticized self-indulgence, Sinclair's Beckton collapses the distinction between subject and object, which in critiques of ruin porn allows the 'othering' of the working class. Its obscenity lies instead in the loss of any proper distance separating viewer and viewed and, therein, the discovery of the viewer's own redundancy. The 'end of the dustbins of history' marks the birth of the world in which the past is never 'over', it is simply elsewhere; the site of industry is not superseded it is simply displaced to another location within global space, and in this sense Beckton is Huế. As an expression of the globalization of waste, Sinclair's Beckton thus renders the question 'how do we feel about the past?' and its corollary, 'what do we do with the past?', redundant. For here the persistence of garbage itself becomes an answer. 'We' cannot do anything with the past; we can no longer reduce garbage to that which we reject and which consequently produces us as subjects of modernity, who become who we are through what we reject.

In attempting to make sense of the function of the garbage produced in this globalized context, it is instructive to return once more to Marx, but this time, rather than his theorizations of value, it is his notion of the General Intellect outlined in the *Grundrisse*'s 'fragment on machines' that seems more helpful. Here Marx argues that at any given moment in history the sum of human knowledge is embodied in its technology. He also develops the argument that because capital will always prefer fixed costs (represented by machines) to the variable costs of living labour, living labour will come increasingly to find itself as a 'living accessory' (1993: 694), a supplement to its own intellect. It is an argument that seems to take on considerable resonance within a globalized economy where precarity and productive obsolescence become the lived experience of workers around the globe and where the automation of ever more forms of production has become the dominant technological narrative, from 3D printers to driverless cars to drone-delivered groceries to massive open online courses (MOOCS). Within the context of this generalized redundancy, Beckton takes on its final significance as a reminder that within a global economy the most revelatory form of the General Intellect might be found not in the latest technology, but within the contemplation of the forms and spaces produced by global garbage.

Note

1 The film which accompanies three tracks from The Smiths is formally attributed to Derek Jarman, but was compiled by Jarman along with John Maybury, Richard Heslop, and Chris Hughes. According to producer James MacKay, Heslop edited the 'The Queen is dead' section, Maybury 'Take me out tonight' and Jarman 'Panic', each working from pooled footage (MacKay 2014).

14 Disposable architecture – reinterpreting ruins in the age of globalization
The case of Beirut

Judith Naeff

Introduction

This chapter engages with those modern ruins for which no final decision on preservation, restoration, or demolition has been made. I will argue that this uncertain status imbues the ruins with a significance that has been overlooked in most of the academic literature on modern or contemporary ruins. The condition of possible impending demolition is important because it resonates with the pre-carization of life in the era of globalization. Within the context of a volume on global garbage, one might expect a reading of ruins that addresses its state of dereliction and decay. However, approaches that focus on the materiality, affect, and aesthetics of ruins (e.g. Edensor 2005) leave little room for the politics of the past and present that have conditioned these structures. Instead, I will use the concept of disposability, which is equally related to the dynamics of waste, but which points more clearly towards forms of violence that are operative in the production and management of waste.

The case of contemporary Beirut is illustrative, because some of its civil-war ruins are still standing amidst a frenzy of highly speculative real estate development, in a historical period punctuated by episodes of war, assassinations, and street clashes. The discussions revolving around these ruins focus mostly on their significance as bearers of memory. I will argue that the fascination with these structures equally arises from the improbable reality that they are still standing as derelict giants in a violently transforming cityscape. This has become particularly meaningful, because the uncertainty of their fate has become interwoven with the uncertainty of the fate of Beirut itself.

However, the argument is relevant to many other cities across the world. The global spread of risk on the one hand (e.g. Beck 2009) and the global precarization of labour on the other (e.g. Gray 2004) have increased the awareness of a shared vulnerability of all (e.g. Puar 2012). The destruction of architecture is an important phenomenon in this respect. Architecture is constructed first and foremost to protect and give shelter and, second, to order space in meaningful ways. Moreover architecture is intended to last, to at least outlive its makers and to give an impression of durability. Its destruction unequivocally reveals the exposure of our bodies to violence when meaningful structures are shattered – when our shelter, or even our world, is threatened.

The violent destruction of architecture has become structural in many places. For example, the structural destruction of domestic architecture by the Israeli Defence Force in Gaza, US drone attacks in Pakistan and the clearances of slums by the governments of Malaysia, Zimbabwe, Brazil, and India, among other countries, put the inhabitants of these areas in an acute state of permanent anxiety and utter precariousness. In other cases, destruction may not be structural, but imagery of these more exceptional forms of destruction has become part of our daily lives. The effects of unintended and natural disasters, such as floods, earthquakes, and oil spills, for example, are increasingly well registered and increasingly widely circulated, with amateur footage bringing various inside perspectives into our homes. Within a larger discourse of global warming, population growth, and financial crisis, an increase in catastrophes is commonly perceived or anticipated. Finally, the spectacular destruction of modern architectural pride on 9/11, of which images have proliferated across the globe, has revealed the vulnerability of all, even of what was considered to be the centre of power in the world.

At the same time, neo-liberal urban renewal schemes from Beijing to Johannesburg, and from London to Istanbul, constitute another form of urban violence. If all urban renewal projects entail the violent destruction of the old, its aim in the neo-liberal age has shifted from the regulation of urban populations to the accumulation of capital. This means that 'the new' of neo-liberal urban renewal is especially unwelcoming to the poor, whose condition in these cities has become increasingly precarious. Although the two forms of violence are different – the first with the stated intention of destruction, the other with the stated intention of construction – it is important to be aware of two consequences that they share: the violent transformation of the built environment and the precarization of certain populations. Moreover, various scholars have investigated how the two forms of violence intersect (e.g. Graham 2004; Chatterjee 2009).

Starting from an analysis of civil-war ruins in contemporary Beirut, this chapter seeks to complement readings of ruins that rely on memory studies with a reading that relies on the concept of disposability. After sketching the social and historical context of contestations over Beirut's ruins, I will discuss some of Beirut's most iconic ruins and the way in which they feature in contemporary artistic practices. Subsequently, I will briefly outline the traditional reading of these and other ruins as structures that refer to the past. Then I will introduce and elaborate the concept of disposability in relation to the notion of precariousness, building on the works of Giorgio Agamben and Judith Butler. An analysis of two works by the Lebanese artist Rayyane Tabet, finally, will demonstrate how the disposable state of Beirut's ruins resonates with a more broadly shared experience of living in Beirut today – an experience of precarious provisionality that arguably reflects global transformations in contemporary urban life and culture.

Architecture and memory in Beirut

The ruins that will be discussed in this chapter are modern structures that have been damaged relatively recently and have since been subject to gradual decay. These ruins were all damaged during Lebanon's civil wars (1975–1990). However, the concerns about the fate of architectural heritage that continues to be uncovered during construction-site digging reveal that my argument is not exclusive to those ruins (e.g. Naccache 1998; Sader 1998). The buildings are generally abandoned, but some may have been used for other purposes than they were intended for, especially during and directly following the war, by militiamen, squatting refugees, or, in the case of the Holiday Inn hotel, by the Lebanese army. The most important characteristic for my argument is that the structures have lost their intended function and that no maintenance, preservation, restoration, or demolition is taking place or being planned.

This means that a ruin like the characteristic 'yellow house' falls outside the considerations of this chapter. This ruin of distinct neo-Ottoman architecture, the walls of which are covered with bullet holes and graffiti, is one of the rare cases where a preservation strategy has been implemented. The structure will be largely preserved in its current state of ruination and a new part will be added in order to open up *Beit Beirut*, which includes a memory museum, cultural centre, auditorium, and cafeteria. This is a first for the city of Beirut, and it remains to be seen how an institutionalized memory museum will affect the debates on memory and heritage in the post-conflict city. To understand these debates and the symbolic value of Beirut's ruins, it is imperative to provide some socio-historical context.

After the civil war was ended with the Taif agreement in 1990, a broad economic reconstruction plan was developed. Real estate development played a major role in the plan to get the disastrous economy back on its feet (Baz 1998). The redevelopment of Beirut's Central District has been the crown piece of this reconstruction plan. Before the war, the district housed the souks, the fancy seaside hotel district, the banks, a transportation hub, shops, cinemas, and cafés. It was the district where Beirutis of all communities mingled in their shopping and leisure, and where tourists were welcomed in luxury leisure facilities. At the outbreak of the civil war in 1975, the district was quickly seized and looted, partly because it was one of the few districts that had no clear political and religious identity, partly because the high rises of the hotels were strategic sniper spots from which fighters could control access to the port, and partly because a seizure would mean a symbolic victory over the bourgeois character of the area (Fregonese 2009). It became the northern end of the demarcation line that separated the city between a leftist pro-Palestinian and predominantly Muslim west and a right-wing anti-Palestinian east of predominantly Christian liberal democrats. As a frontline, the district was severely damaged throughout the long years of fighting and eventually turned into an abandoned wasteland where stray dogs roamed the overgrown streets.

After the war, it was generally agreed upon that fast reconstruction of this area had to be prioritized as a symbol of national unity. Since the governmental institutions needed for such a massive development project were partly non-existent, and partly bankrupt, understaffed, and dysfunctional, the law was adjusted, allowing a private company, the Lebanese Society for the Development and Reconstruction of Beirut's Central District – better known under its French acronym Solidere – to appropriate all properties in the area allotted to them. The expropriated owners received shares in the company in return.

Because the concrete architectural questions of reconstruction provided a platform to discuss vital issues of memory and identity without naming and blaming, and because the central district's symbolic value as a common ground, Solidere's project formed the incentive for what has been described as 'the first public debate since the beginning of the war, and the first on urban matters in Lebanon's history' (Beyhum 1992: 50). The beginning of Solidere's reconstruction project involved the large-scale demolition of the remaining buildings, many of which could have been restored. Within this framework, the 'memory debate' became a central topic. Critics argued that the contractors were indifferent to important pieces of heritage in favour of global capital, denying the city its rich history (Makdisi 1997; Sader 1998; El-Khoury 1998). The spatial erasure of old structures was identified as part of what came to be called collective, or official, amnesia, referring to the deliberate disengagement with recent history on the part of the authorities. The general amnesty law that was issued in 1991 meant that nearly all post-war politicians were old warlords and that the conflicts that had sparked the violence remained unresolved. As a result, the government had no interest in clearing up the fate of the estimated 17,000 disappeared, or other past crimes. Additionally, the continuing divisions that split both leadership and people over crucial historical events and the nature of national identity has prevented the formulation of a unified national history to be taught in Lebanese schools. The relentless rounds of demolition in the city centre were understood as a material expression of the will to cover up Beirut's divided and divisive past.

Solidere's marketing strategy seemed to confirm this interpretation. The company has promoted an image of Beirut's city centre that combines the myth of a glorious past dating back to Phoenician times, a popular nostalgia for pre-war prosperity and *joie de vivre*, and a vision of a future competitive metropolis in a globalizing world. Some have argued that 'Solidere's slogan "Beirut, ancient city for the future" comfortably leaves out the present and immediate past' (Haugbolle 2010: 86; El-Dahdah 1998). A group of intellectuals and activists have been urging the company and the authorities to create a space that would engage with Lebanon's troubled recent history. As 'memory makers' this group has framed their argument in a didactic formulation – 'we should remember so as to learn from our mistakes' – as well as in a psychoanalytic formulation – 'we should process our memories in order for our wounds to heal' (Haugbolle 2010: 64–95).

Contestations over the fate of large civil-war ruins should be seen in the context of these debates. Those who argue for preservation point out that the ruins have become distinctive landmarks in Beirut's cityscape that testify to its complex and troubled history. Their presence furthermore functions as an admonition, arguably working against a collapse into renewed civil war, something which most Lebanese continue to fear. Opponents underline the lack of function and efficiency to argue that preserving ruins is a waste of space. Their lack of use value constitutes a waste of the high land-value in the area. Those in favour of demolition mostly argue in a market logic, which views demolition as 'creative destruction' necessary for the stimulation of economic development. However, some refer to the symbolic value of ruins when arguing for demolition. They point out that these buildings often invoke traumatic memories of violence and suffering. Therefore they cause unnecessary pain. Moreover, institutionalized preservation could potentially turn the ruin into a spectacle befitting the global heritage industry, which would inappropriately turn memories of suffering into a commodity.

Beirut's iconic ruins

Some of the most dramatic ruins have attained a striking presence in the literary, cinematic, and artistic imagination of post-civil-war Beirut. The *Murr* tower, for example, is a giant concrete structure that was supposed to house the world trade centre but construction of which was halted when the civil war broke out. The tallest tower of the city until the mushrooming high rises of the 2000s, this building is charged with symbolic meaning. As a world trade centre to be, the *Murr* tower symbolizes the pre-war commitment to capitalist modernity as a project that was prematurely and violently aborted, reminiscent of the myth of Babel. In addition, the tower was an infamous sniper spot and site of torture during the war, which adds a traumatic layer to the signification of the place, connoting unspeakable cruelty and violence.

In the aftermath of Israel's war on Lebanon in 2006, two films refer to the *Murr* tower as a valuable war monument. In Ghassan Salhab's short video *(posthume)*, the tower is appreciatively called a 'witness to all our wars' (Salhab 2007). In the feature film *Je veux voir* (Hadjithomas and Joreige 2008), main character Rabih Mroué tells Catherine Deneuve that he thinks the building should stay, as a reminder of the war. In Mroué's own play *How Nancy Wished that Everything Was Just an April Fool's Joke* (Mroué and Toufic 2007), the top of the *Murr* tower is the setting for an unlikely meeting between four old fighters. Written at a particularly tense period in Lebanese politics, Mroué explains that '[the piece] came out of a real fear of another civil war. For those of us who already lived through one, we can live through almost anything – but not that, not again' (cited in Wilson-Goldie 2007). The characters of the play have all come to claim the strategic building in anticipation of a new civil war. Recounting their personal histories in which they repeatedly die and then resume fighting, and in which they repeatedly switch allies – an allusion to the political manoeuvring of Lebanon's political parties – the play is a powerful denouncement of war and its utter uselessness. The tower is thus appropriated in the function of admonition.

Figure 14.1 The *Murr* tower (photograph by the author, 2011).

Another example is the Holiday Inn, a luxurious hotel with a revolving restaurant on the top floor, opened only months before the war broke out. It was quickly seized during the infamous Battle of the Hotels (1975–1976). While the Phoenicia hotel has been restored to its former splendour, the owners of the Holiday Inn Hotel have been locked in dispute for decades, leaving the fate of the imposing structure undecided. In 2014 a public announcement declared that the structure was sold to a new owner, but no development plan has been revealed at the moment of writing.

The ruin is famous for its size and for its seemingly ostentatious display of the traces of the legendary battle amidst the new high rises of Solidere's shiny district. Grasses can be seen growing on the open floors of the building and all sides are pockmarked by bullet holes. One side reveals an impressive crater of a larger mortar impact. These scars testify to a legendary episode of the civil war. The Battle of the Hotels was both strategic and symbolic. Factions fought bitterly over the seizure of the hotels because they provided control over access to the port. For this reason, the ruin has been used in the post-war period by the Lebanese army to store vehicles and materiel, symbolically reclaiming power. In the heart of Beirut's pre-war decadence, the vertical modern architecture proved a spectacular setting to the scenes of violence. Raed Yassin is an artist that has focused on this aspect. In his series *China* (2012a), legendary battles of the Lebanese civil war are depicted in miniature style on Chinese vases. One of the vases depicts the Battle of the Hotels. This approach removes the battle from the realm of direct memory and trauma, to the realm of myth and legend, as well as to a globalized industry of art production and circulation.

In contrast, Ayman Baalbaki has painted the Holiday Inn ruin in thick layers of oil paint. Some of the various canvases dedicated to the structure bear titles such as 'the sniper' or 'seeking the heights', referring to the building's pasts. One critic remarked that Baalbaki's use of the

> impasto technique that uses a painting knife to carve up the layers of the canvas [– e]ven though the impressionists used that style for enhancing colour and light – [...] is used here with dark colours to amplify the layers of destruction and fragility of the urban fabric of Lebanon.
>
> (Sabounchi 2010)

This interpretation becomes even more pertinent when viewed within the context of his complete oeuvre. Significantly, Baalbaki painted several series of urban landscapes in a similar style suggesting thematic relations between the different environments. His paintings of the tower of Babel as imagined by Pieter Bruegel de Oude (2006b) prefigure and strongly resemble his depictions of the destruction brought about by Israel's 2006 bombings in Southern Beirut (the Tammouz series, 2006a). Brighter but close in style are his earlier impressions of the neighbourhood Wadi Abu Jamil (such as the four panels, named *Ciel Chargé de Fleurs*, 2004), where the painter lived after his family had fled their home village, and which has now been violently transformed into Solidere's upscale 'ancient city for the future'.

Throughout these various themes, Baalbaki has painted not only on canvas, but also on textile printed with colourful flowers of the kind village women wear in Southern Lebanon. The flowers that are visible through the strikes of paint of the sky, are reminiscent of the humanity and dignity of the people caught up in the continuing history of violence and displacement, without relapsing into essentialist ideas of authenticity or tradition, because, Baalbaki points out, these fabrics were 'quintessentially postcolonial' and manufactured in China (Wilson-Goldie 2006). Moreover, his installations of bundles wrapped up with colourful pieces of cloth and textile and tied up with ropes express his engagement with the condition of displacement that continues to mark Lebanon's identity. So Yassin interprets the Holiday Inn as a referent to the past, a legendary aberration that he desires to 'literally take outside this acute mode of interiority', to give it 'a sense of protracted closure in space' by petrifying 'a geopolitical and personal constellation of experience [...] into a thing' (Yassin 2012b: 74, 72). Baalbaki, in contrast, positions it in an ongoing history of violence and destruction that continues to affect the daily lives of people in Lebanon and elsewhere, drawing parallels between the mythical destruction of Babel and the people of Lot and the destruction of southern Beirut in the summer of 2006, between the iconic civil-war ruins of the Holiday Inn and the Burj El Murr, and the history of displacement of his own family, as well as that of a neighbourhood such as Wadi Abu Jamil (more on the continuities in Baalbaki's work in Rabbath 2009 and Tamraz 2009).

Finally, the dome, or egg, is a former cinema called City Palace. Its striking round architecture and ripped-open armed concrete contrast sharply with the vertical architecture and impeccable newness of neighbouring buildings. In terms of memory, the site is important because cinemas play a very important role in nostalgic imaginaries of the pre-war city. But the dome is particularly appreciated for its unruliness, in contrast to the concrete rectangular shapes of the *Murr* tower, which match the geometry of much of the surrounding architecture. In Rabih Jaber's novel *Beyritus, Underground City* (2006), the main character is a security guard of the dome, making him fully part of the neo-liberal logic of Solidere. When he is chasing an intruder inside the ruin, he falls into a pit and discovers a hidden underground city. In his yearlong stay underground, the labyrinthine city evokes a variety of repressed memories, while offering a glimpse of an alternative society, albeit one that is marked by intense suffering. In its unruliness, the site resists the logic of capitalist urban space and allows the present city of Beirut to connect to both its past and its unconscious or repressed other in ways that commoditized heritage could not.

In Mazen Kerbaj's graphic novel *Lettre à la Mère* (2013), sec black-and-white drawings of the ruins of Beirut form the main background of his letter to his irresponsible mother: the city of Beirut. Having abandoned him at birth, the mother always escapes him, even though she is always omnipresent. Her disengagement is always because of the war, but while we mortals are ageing, says Kerbaj, she keeps renewing herself, 'marching over our cadavers with shoes of concrete' (2013: n.p.). However, in a later section in the book, he returns to the dome. This time, the ruin is not rendered in black pencil, but in photographs that are illustrated

Figure 14.2 Holiday Inn, Ayman Baalbaki, 2011 (courtesy of the artist).

with colour. The text explains that the dome was one of the few structures outside east Beirut that he could see from his grandparents' place, before the end of the civil war, and that he had always imagined it to be a UFO. As a place that stimulates the imagination, the UFO later invited creative uses. He visited a rave party at the ruin in 1995 and performed experimental music that incorporated sounds of the Israeli bombs of 2006 to a horrified public in 2007. He concludes that he has not

Figure 14.3 The dome, interpreted as UFO, Mazen Kerbaj, 2013 (courtesy of the artist).

found 'a symbol that is more maternal than my UFO-egg' (Kerbaj 2013: n.p.). However, its imminent disappearance, announced for over a decade, does not bother him, because 'in the worst case, it will be replaced by a glass tower that will make for a super-UFO at the end of the next war' (2013: n.p.).

From ʾAṭlāl to disposables

Against the backdrop of these contestations over urban planning, and bearing in mind the various ways in which ruins have inspired artists and intellectuals alike, it becomes imperative to look at what the ruin signifies. In his *Expositions*, Philippe Hamon points out that the 'semantic complexity' (1992: 59) of the ruin lies in its dual significance. Its temporal component provokes 'the pure effect of memory' (Hamon 1992: 62), its spatial component 'unveils and exhibits its structure' (Hamon 1992: 60). This second aspect has inspired many thinkers, from Walter Benjamin to various post-structuralists, to use the ruin as a concept that in its material porosity, slippage, and decomposition allows for the indefinite postponement of closure and wholeness of meaning in general. If the ruin's spatial peculiarity makes it a privileged symbol and metaphor, debates revolving around the fate of actually existing ruins rarely mention it. Instead, they refer mostly to the ruin's temporality.

In the Arab world, the ruin's effect of memory is strongly rooted in the literary motif of *wuqūf ʿalā al-ʾaṭlāl*, 'stopping before the ruins'. In Imruʾul-Qays's pre-Islamic *muʿallaqa*, a monumental ode that is a fundamental part of the Arabic educational curriculum, the narrator sets himself weeping at the traces, *ʾaṭlāl*, of the departed Bedouin encampment of his beloved. The traces materialize the irrecuperable loss of the loved one. It not only sets off a train of memories but also embraces and aestheticizes non-teleological longing. Today still, the ruin motif is a literary theme in novels, poems, and songs, closely associated with nostalgic yearning (Kilpatrick 2000).

In short, the ruin represents 'the presence of an absence, [...] the ruin functions as a sort of negative punctuation of space' (Hamon 1992: 62). However, the ruin's symbolic value goes beyond mere negation. It embodies not merely absence, but the ongoing process of ruination, the inevitable passing of time, of peoples, and of civilizations. As such, the ruin can be a *vanitas* symbol. What Andreas Huyssen calls the authentic ruin of modernity is marked by a 'consciousness of the transitoriness of all greatness and power, the warning of imperial hubris, and the remembrance of nature in all culture' (Huyssen 2010: 21). These authentic ruins reveal the ambiguity that has always been inherent to modernity. They are structures on which modernity's 'fear of and obsession with the passing of time' are projected (Huyssen 2010: 19).

However, unlike the traces of an encampment, or the gradual decay of past glory, Beirut's architectural ruins imply violent destruction. They do not make us aware of the existential passing of time and the natural decay of great achievements, but of violence and destruction that is political. Consequently, the memories that they invoke do not come in the form of merely nostalgic yearning, but in the form of painful memories of violence. Ken Seigneurie's discussion of ruins in contemporary Lebanese literary fiction claims that '[f]rom the contemplation of ruins emerges a yearning for the dignity *that should have been*', rather than a yearning for an idealized past (2008: 59, emphasis in original).

Huyssen claims that for authentic ruins there is no place in late capitalism's culture of commodity and memory. Ruined structures are restored and preserved until they resemble pre-packaged commodities, or become obsolete and are quickly disposed of. Greg Kennedy, in his study of trash, points out that our culture of disposables relies heavily on the carelessness of human beings. He defines trash as those objects to which reason attributes value but which in practice we nevertheless waste. Therefore, trash is disturbing, because it reveals how we act against our own reason (Kennedy 2007: 5). To place carelessness at the centre of late capitalism's culture of disposables is important, because it implies an exposure to violence – of objects considered disposable – that is socially and politically constituted.

Such an exposure to violence resonates with what Giorgio Agamben has conceptualized as 'bare life'. It is life on the edge of life, abandoned by any form of protection and subsequently left to mere survival. From the Roman *homo sacer*, to the medieval bandit, to the concentration camp inmate and the contemporary refugee, it is through the exclusion from law – rather than punishment by law – that sovereign power over life and death is exerted (Agamben 1998). Although Agamben's theory has shown an inappropriate disregard for historical specificity, he has nevertheless offered the important insight that the unequal exposure to death is central to contemporary politics (see also Noys 2005).

I would like to move from the narrow notion of bare life as life exposed to death to precariousness as life exposed to any kind of harm. Even if a subject is not in mortal danger, the threat of unemployment, displacement, detainment, physical pain, and personal or material loss may produce a precarious condition. Precariousness refers to the vulnerability of human beings to harm. If every human being is precarious insofar as every subject is dependent on others for shelter, sustenance, and care, the precarious condition is not merely existential. Precariousness may be a way of being in the world, but only a way of being in relation to others. Hence, Judith Butler maintains that precariousness is 'a function of our social vulnerability and exposure that is always given some political form' (in Puar 2012: 169). In other words, to expose the other to harm is a political act of violence.

In Lebanon, the repeated instances of violence in the period since the end of the civil war in 1990 point towards a structural exposure to harm caused by the unresolved divisions within Lebanese society and in the larger region. Another form of structural exposure to violence of the urban environment – and its population – was produced by the way post-war governments have tapped into a geo-Darwinian discourse of interurban competition in a globalizing world. These two forms of precariousness are not limited to Beirut – they can be found all over the world. The measure of vulnerability to these forms of violence is unequally distributed. Lower classes, women, immigrant labourers, and refugees are more vulnerable than others. However, the structural exposure to violence of Beirut's geography affects all city dwellers to a certain extent. The various artistic engagements with ruins discussed above reveal the continued engagement with ruins as referents to the past, as well as admonitions for the future. In some cases, however, this has been coupled with an awareness of their continued exposure to violence, that is their disposability, such as in the works of Baalbaki and Kerbaj.

From the disposable to the provisional

To conclude, I would like to discuss the work of an artist that has engaged in various ways with the precariousness of Beirut's urban environment. In response to the request of art magazine *Peeping Tom Digest* to respond to a set of comments and questions, Rayyane Tabet, trained as an architect, reprinted a cover of the *New York Times*. It showed a picture of an endless line of trucks transporting the rubble of bombed-out buildings from the southern suburb of Beirut to a landfill just outside of town in the direct aftermath of the war with Israel in 2006. Tabet added the comment 'if I were to write a text on sculpture, it would probably begin with this image' (Tabet 2013). This comment can only be fully understood in the context of his art.

For his installation *1989* (2012a), he meticulously cut out the window frame and door of his old childhood bedroom. He had them shipped to the exhibition space where he incorporated them in a large canvas structure. The installation reminded of a tent, the ultimate architectural form of transitoriness. A short story, written by the artist together with John Greenberg, was handed to visitors of the exhibition. The story tells how Rami one night gets up to fetch a glass of water and upon return finds out his bedroom has disappeared. When his parents and neighbour respond utterly laconically to the news, continuing their everyday lives around the gaping hole in the building, Rami runs away. Outside, a soldier stops him.

> 'Hey, you!' the soldier said, addressing Rami as he would a common street urchin. 'From whom and to where are you running?'
>
> 'From no one and to nowhere,' Rami said. 'It's just…' – Rami pointed to the prominent hole in his building's façade – 'it's just that my bedroom and everything in it have vanished.'
>
> The soldier laughed, not mockingly, but also not innocently. He laughed desperately and gestured widely, saying, 'Look around you, boy. Look at all the places that have vanished and tell me what you see.'
>
> (Tabet and Greenberg 2012)

Figure 14.4 1989, Rayyane Tabet, 2012 (courtesy of the artist).

Disappearing architecture is thus posited not as a rupture, an anomaly, but as a defining condition of the contemporary city of Beirut. Rami's coming of age is set off by the loss of his bedroom. It is his realization of the utter fragility of the structures that founded his lifeworld that constitutes his loss of innocence. The fact that the adults surrounding him consider this radical instability as nothing out of the ordinary allows for such an interpretation. Moreover, the storyline follows the classical trope of a journey that brings about the transition to a new stage, a new responsibility. In the last two paragraphs, the tent-like structure of the installation attains new meanings:

> Beyond the empty floor, someone had tacked a sheet across the open hole in the building's façade. The sheet billowed and snapped, punching convex into the apartment before being sucked outside in the vacuum of wind. [...] He imagined the feel of the chilly tile in the summer morning against his bare feet. He smelled the freshness of his linens against the musty odor of his laundry basket. He felt the burst of warmth and blinding light when his mother pulled the curtains aside to wake him in the morning. Rami allowed the memories to play across his mind, and he willed them to fill the void and animate the billowing sheet in front of him. Convinced that he would wake up in bed, Rami Katchadouri succumbed to the innocuous nothing of sleep.
>
> (Tabet and Greenberg 2012)

The installation resembles the temporary sheet that is unable to cover the void that plagues Rami, and arguably life in Lebanon more generally. It is an aesthetically appealing installation, allowing Rami to try to fill its emptiness with memories of a secure and joyful childhood. He clings to the past in a stubborn conviction of an imminent return. Rami is thus incapable of accepting the harsh reality of his precarious present. The position of the artist is less obvious. The sheer beauty of the installation and some delight in the absurd that shimmers through the short story suggest a more ambiguous attitude.

His *Architecture Lessons* (Tabet 2012b), shows this ambiguity more clearly. In this work, Tabet has recast his old wooden toy blocks in large quantities in grey concrete. The exhibition room houses endless combinations of the blocks forming a stretched-out concrete city. A video shows the original blocks of wood in primary colours with two hands permanently constructing and deconstructing different structures in a loop. Here, a much less personal geography of precarity is performed – one in which there is room for the architect to rejoice in the space for creativity opened up by instability and provisionality.

In both works there is a sense of exposure to harm as a condition rather than a historical event that is a passing deviation. Going back to his contribution to *Peeping Tom's Digest*, it becomes clear that Tabet reads the destroyed buildings on their way to the landfill as the general condition of contemporary Lebanon, which is in a constant process of decomposition and reconfiguration, and within which sculpture in Lebanon today needs to find its place. While the artist may be well aware of, and concerned with, the suffering of individuals, such an approach

Figure 14.5 Architecture Lessons, Rayyane Tabet, 2012 (courtesy of the artist).

allows him to recognize that destruction can be creative. It resonates with Kerbaj's depiction of the dome. The structure allows for experimental events, and for the imagination of a UFO landed in the centre of Beirut, only because it is provisional. He cannot lament its imminent ending, because it has been defined by its imminent ending all along and because the structural precariousness of architecture in Beirut will inevitably affect whatever will be built in its place.

In conclusion, contemporary ruins whose fate have remained undecided have various layers of meaning. As structures that have lost their intended function, they signify the absence of what is no longer there, and of what could have been. Produced by intentional destruction, the ruins in this chapter obviously bear witness of past violence. This is not only the case in post-war settings. The industrial ruins of northern England and Detroit not only signify the past of triumphant industrialism, or even the passing of all great empires, but also point to the violence with which a transforming capitalism has disposed of these structures, and to the suffering of those who were exposed to that violence. Looking at artistic engagements with these structures, issues such as memory, history, and trauma form important layers of meaning. One way in which this meaning is directed towards the future, is in the argument of admonition. The play of Mroué and Toufic, set on top of the *Murr* tower, forms an example that shows past experience being projected onto the present and immediate future to perform a cautionary function.

In this chapter, I have argued that within the context of late capitalism's culture of commodities and disposables, another layer of meaning has potentially become equally important. Highly speculative urban redevelopment projects across the globe leave little room for ruins whose heritage function has not been commoditized. The disposable condition of these ruins resonates with the precarious condition of life in Beirut. The ruins are exposed to the imminent threat of violent bulldozers. Beirut is exposed to the imminent threat of renewed war. The utter precariousness of both life and architecture in Beirut is structural. This may be particularly obvious in Beirut, but it is definitely not limited to Beirut. The global circulation of money, goods, and labour; the blurring of work and leisure time; global warfare and terrorism; the incorporation of domains of care – such as welfare, health, and education – into the capitalist market and its concomitant careless culture of commodities and disposables; they have all violently produced an increased exposure to various forms of harm and displacement on a global scale.

To appreciate the ruin as a disposable structure is important for two reasons. First, it removes both its past abandonment and its imminent disappearance from the existential realm of *vanitas*, death, and the passing of time, putting it back into the political realm of an exposure to harm that is conditioned by power structures. Second, approaching disposability as a condition, rather than as a transitory historical anomaly, allows for an appreciation of the provisional as a condition that opens up creative possibilities, even despite an acute awareness of injustice and suffering.

References

Abbas, A. (1997) *Hong Kong: Culture and the Politics of Disappearance*. Minneapolis, MN: University of Minnesota Press.

Abujidi, N. (2014) *Urbicide in Palestine: Spaces of Oppression and Resilience.* New York, NY: Routledge.

Adelkhah, F. (1999) *Being Modern in Iran.* London: Hurst & Company.

Adorno, T. (1973) *Negative Dialectics*, trans. E.B. Ashton. London: Routledge and Kegan Paul.

Adorno, T. (1999) *Aesthetic Theory.* London: Athlone.

Agamben, G. (1998) *Homo Sacer: Sovereign Power and Bare Life*, trans. D. Heller-Roazen. Stanford, CA: Stanford University Press.

Albanese, D. (2014) 'Interview With Art Is Trash'. *The Dusty Rebel*, 27 June, available at: www.thedustyrebel.com/post/89968015742/interview-with-art-is-trash (accessed 4 September 2015).

Alexander, C. and J. Reno (2012) *Economies of Recycling: The Global Transformation of Materials and Social Relations.* London: Zed.

Allen, A. (2013a) 'Shifting Perspectives on Marginal Bodies and Spaces: Relationality, Power, and Social Difference in the Documentary Films *Babilônia 2000* and *Estamira*'. *Bulletin of Latin American Research* 32(1): 78–93.

Allen, A. (2013b) 'Sites of Transformation: Urban Space and Social Difference in Contemporary Brazilian Visual Culture'. PhD dissertation. Cambridge: University of Cambridge.

Allen, C. (2007) 'Of Urban Entrepreneurs or 24-hour Party People? City-Centre Living in Manchester, England'. *Environment and Planning* A 39(3): 666–83.

Allen, C.J. and J.M. Hamilton (2010) 'Normalcy and Foreign News'. *Journalism Studies* 11(5): 634–49.

Allitt, P. (2014) *A Climate of Crisis: America in the Age of Environmentalism.* New York, NY: Penguin.

Allman, J.M. and V.B. Tashjian (2000) *'I Will Not Eat Stone': A Women's History of Colonial Asante*. Oxford: James Currey.

Allmendinger, P. and G. Haughton (2012) 'The Fluid Scales and Scope of UK Spatial Planning'. *Environment and Planning* A 39(6): 89–103.

Allum, F. and P. Allum (2008) 'Revisiting Naples: Clientelism and Organized Crime'. *Journal of Modern Italian Studies* 13(3): 340–65.

Allum, P. (1973) *Politics and Society in Post-War Naples.* Cambridge: Cambridge University Press.

Almeida, M. (2006) 'Estamira: A Salvação no Lixo'. *Digestivo Cultural*, available at: www.digestivocultural.com/colunistas/coluna.asp?codigo=2058&titulo=Estamira:_a_salvacao_no_lixo (accessed 20 October 2014).

Amin, A. (2012) *Land of Strangers*. Cambridge: Polity Press.

Amir-Ebrahimi, M. (2006) 'Conquering Enclosed Public Spaces'. *Cities* 23(6): 455–61.

Ammons, A.R. (2002/1993) *Garbage*. New York, NY: W.W. Norton & Co.

Anonymous (n.d.a) 'Anonymous Memoires of a Trading Post at Sapele'. United Africa Company (Unilever Archives and Records), 1/11/14/3/1.

Anonymous (n.d.b) 'Anonymous Memoires of a Trading Post at Sapele'. United Africa Company (Unilever Archives and Records), 1/11/14/3/1b.

Appadurai, A. (1996) *Modernity al Large: Cultural Dimensions of Globalization.* London and Minneapolis, MN: University of Minneapolis Press.

Appleyard, D., K. Lynch, and J.R. Myer (1964) *The View From the Road.* Cambridge, MA and London: MIT Press.

Armiero, M. and G. D'Alisa (2012) 'Rights of Resistance: The Garbage Struggles for Environmental Justice in Campania, Italy'. *Capitalism Nature Socialism* 23(4): 52–68.

Armstrong, H. (2006) 'Time, Dereliction and Beauty: An Argument for 'Landscapes of Contempt'. Presented at The Landscape Architect, IFLA Conference.

Assise di Napoli (2007) 'Comunicato stampa', 20 March, available at: www.napoliassise.it/comunconprog.pdf (accessed 10 October 2014).

Augé, M. (2009) *Non-Places: Introduction to an Anthropology of Supermodernity*, trans. J. Howe. New edition. London and New York, NY: Verso.

Austin, D. (2012) *Forgotten Landmarks of Detroit.* Stroud: The History Press.

Austin, D. and S. Doerr (2010) *Lost Detroit: Stories Behind the Motor City's Majestic Ruins.* Gloucester: The History Press.

Baalbaki, A. (2004) *Ciel Chargé de Fleurs.* Acrylic on printed fabric laid on wood. 180 × 180 cm.

Baalbaki, A. (2006a) *Tammouz.* Series of acrylic on canvas. 42 × 30 cm.

Baalbaki, A. (2006b) *Tour de Babel.* Series of acrylic on canvas. Various sizes.

Baalbaki, A. (2010) *Holiday Inn Hotel: Seeking the Heights.* Oil on upholstery printed fabric. 189 × 199 cm.

Baalbaki, A. (2011) *Burj El Murr.* Acrylic on printed fabric laid on canvas. 200 × 150 cm.

Baalbaki, A. (2011) *Holiday Inn.* Acrylic on printed fabric laid on canvas. 205 × 155 cm.

Babcock, B. (1978) *The Reversible World: Symbolic Inversion in Art and Society.* Ithaca, NY: Cornell University Press.

Babenco, H. (dir.) (1982) *Pixote.*

Baines, C. (2012) 'The Wild Side of Town', in A. Jorgensen and R. Keenan (eds), *Urban Wildscapes*. Abingdon and New York, NY: Routledge, xii–xv.

Bal, M. (2003) 'Visual Essentialism and the Object of Visual Culture'. *Journal of Visual Culture* 2(1): 5–32.

Ball, M., C. Lizieri, and B.D. MacGregor (1998) *The Economics of Commercial Property Markets*. London and New York, NY: Routledge.

Ballard, J.G. (1973) *Concrete Island.* New York, NY: Picador.

Barber, K. (1987) 'Popular Arts in Africa'. *African Studies Review* 30(3): 1–78.

Barr, J. (1969) *Derelict Britain.* Harmondsworth: Penguin.

Barron, P. (2014) 'Introduction: At the Edge of the Pale', in: M. Mariani and P. Barron (eds), *Terrain Vague: Interstices at the Edge of the Pale.* London and New York, NY: Routledge, 1–23.

Bataille, G. (1985) *Visions of Excess: Selected Writings 1927–1939*, ed. and trans. Allan Stoekl. Manchester: Manchester University Press.

Bataille, G. (1991) *The Accursed Share, Vol. I*, New York, NY: Zone Books.

Bates, B. (2012) *The Making of Black Detroit in the Age of Henry Ford.* Chapel Hill, NC: University of North Carolina Press.

Baudrillard, J. (1994a) 'The System of Collecting', in J. Elsner and R. Cardinal (eds), *The Cultures of Collecting.* Cambridge, MA: Harvard University Press.

Baudrillard, J. (1994b) *The Illusion of the End.* Cambridge: Polity.

Baudrillard, J. (1995) *The Gulf War Did Not Take Place.* Bloomington, IN: Indiana University Press.

Baudrillard, J. (2003) *Passwords.* London: Verso.

Bauman, Z. (2004) *Wasted Lives: Modernity and Its Outcasts.* Oxford: Polity.

Bayat, A. (2010) 'Tehran: Paradox City'. *New Left Review* 66: 99–122.

Baz, F. (1998) 'The Macroeconomic Basis of Reconstruction', in P. Rowe and H. Sarkis (eds), *Projecting Beirut: Episodes in the Construction and Reconstruction of a Modern City.* Munich: Prestel Verlag, 165–72.

BDP (Building Design Partnership) (1989) 'The Waterways Guide'. Manchester: BDP.

Beck, U. (2009) *World at Risk*, trans. C. Cronin. Cambridge and Malden, MA: Polity Press.

Benjamin, W. (1992) *Illuminations*, ed. H. Arendt. London: Fontana Press.

Benjamin, W. (2006) 'The Paris of the Second Empire in Baudelaire', in M.W. Jennings (ed.), *The Writer of Modern Life: Essays on Charles Baudelaire.* Cambridge, MA: Harvard University Press, 46–133.

Benjamin, W. (2007) *Illuminations*, trans. H. Zohn, ed. H. Arendt. New York, NY: Schocken Books.

Bennett, J. (2004) 'The Force of Things: Steps Towards an Ecology of Matter'. *Political Theory* 32: 347–72.

Bennett, M.M. and A.V. Seaton (eds) (1996) *Marketing Tourism Products: Concepts, Issues, Cases.* London: International Thomson Business Press.

Berens, C. (2011) *Redeveloping Industrial Sites: A Guide for Architects, Planners, and Developers.* Hoboken, NJ: John Wiley & Sons.

Berger, A. (2006) *Drosscape: Wasting Land in Urban America.* New York, NY: Princeton Architectural Press.

Berman, M. (1988) *All That is Solid Melts into Air: The Experience of Modernity.* New York, NY: Penguin Books.

Berman, M. (1996) 'Falling Towers: City Life After Urbicide', in D. Crow (ed.), *Geography and Identity.* Washington, DC: Maisonneuve Press.

Beyhum, N. (1992) 'The Crisis of Urban Culture: The Three Reconstruction Plans for Beirut'. *The Beirut Review* 4: 43–62.

Blanchot, M. (1987) 'Everyday Speech'. *Yale French Studies* 73: 12–22.

Bogo, J. (2014) 'Exclusive: Inside Google's Quest To Bring Polar Bears To You'. *Popular Science*, 27 February, available at: www.popsci.com/article/technology/exclusive-inside-googles-quest-bring-polar-bears-you (accessed 7 September 2015).

Bond, P. (2002) *Unsustainable South Africa, Environment, Development and Social Protest.* Pietermaritzburg: University of Natal Press.

Bond, R. (2005) *Iain Sinclair.* Cambridge: Salt.

Boo, K. (2014) *Behind the Beautiful Forevers: Life, Death, and Hope in Mumbai 'Undercity'.* New York, NY: Random House.

Booth, R. (2014) 'Inside Billionaires Row: London's Rotting, Derelict Mansions Worth £350m'. *Guardian*, 31 January, available at: www.theguardian.com/society/2014/

jan/31/inside-london-billionaires-row-derelict-mansions-hampstead (accessed 21 September 2014).

Booth, W. (1890) *In Darkest England, and the Way Out.* London: International Headquarters of the Salvation Army.

Born, M., H. Furján, and L. Jencks with P.M. Crosby (2012) *Dirt.* Philadelphia, PA and Cambridge, MA: PennDesign and The MIT Press.

Bourdieu, P. (1993) *The Field of Cultural Production: Essays on Art and Literature.* Cambridge: Polity.

Bowman, A.O. and M.A. Pagano (2004) *Terra Incognita: Vacant Land and Urban Strategies.* American Governance and Public Policy Series. Washington, DC: Georgetown University Press.

Bozonnet, J. (2007) 'Les Mines d'Ordures de la Camorra'. *Le Monde*, 24 April, available at: www.lemonde.fr/europe/article/2007/04/24/italie-les-mines-d-ordures-de-la-camorra_901086_3214.html (accessed 4 September 2015).

Brenner, N. (2012) 'What is Critical Urban Theory?' in N. Brenner, P. Marcuse, and M. Mayer (eds), *Cities for People, Not For Profit: Critical Urban Theory and the Right to the City.* Abingdon: Routledge, 11–23.

Brookhenkel (2009) 'Turning Tables'. *Thing Theory*, 22 June, available at: http://thingtheory2009.wordpress.com/2009/06/22/turning-tables-2/ (accessed 13 October 2014).

Brown, K. and W. Beinart (2013) *African Local Knowledge and Livestock Diseases: Diseases and Treatments in South Africa.* Oxford and Johannesburg: James Currey and Wits University Press.

Bullard, R. (2000) *Dumping in Dixie: Race, Class and Environmental Quality.* Boulder, CO: Westview.

Bürger, P. (1994) *Theory of the Avant-Garde*, trans. Michael Shaw. Minneapolis, MN: University of Minnesota Press.

Burke, T. (1996) *Lifebuoy Men, Lux Women: Commodification, Consumption and Cleanliness in Modern Zimbabwe.* London: Leicester University Press.

Burke-Gaffney, B. (2002) 'Hashima: The Ghost Island'. *Cabinet Magazine* 7, available at: www.cabinetmagazine.org/issues/7/hashima.php (accessed 4 September 2015).

Caffentzis, G. and S. Federici (2014) 'Commons Against and Beyond Capitalism'. *Community Development Journal* 49(Suppl. 1), i92–i105.

Calvino, I. (1997) *Invisible Cities*, trans. William Weaver. London: Vintage.

Camille, M. and A. Rifkin (2001) *Other Objects of Desire: Collectors and Collecting, Queerly.* Oxford: Blackwell.

Campkin, B. (2013) *Remaking London: Decline and Regeneration in Urban Culture.* London and New York, NY: I.B. Tauris.

Campo, D. (2013) *The Accidental Playground: Brooklyn Waterfront Narratives of the Undesigned and Unplanned.* New York, NY: Fordham University Press.

Capone, N. (2013) 'The Assemblies of the City of Naples: A Long Battle to Defend the Landscape and Environment'. *Capitalism Nature Socialism* 24(4): 46–54.

Carmona, M. (2010) 'Contemporary Public Space: Critique and Classification, Part One: Critique'. *Journal of Urban Design* 15(1): 123–48.

Carney, P. and V. Miller (2009) 'Vague Spaces', in A. Jansson and A. Lagerkvist (eds), *Strange Spaces: Geographical Explorations into Mediated Obscurity.* Farnham: Ashgate, 33–55.

Carvalho de Ávila Jacinto, A. (2008) 'Da Exclusão à Construcção do Mundo: Estamira e os Tesouros do Lixo', in *Web-Based Publications of the 9th Brazilian Congress of Fundamental Psychopathology*, 2008, IX. Congresso Brasileiro de Psicopatologia

Fundamental, Universidade Federal Fluminense: Psicopatologia Fundamental, available at: www.psicopatologiafundamental.org/pagina-temas-livres-459 (accessed 24 October 2014).

Castells, M. (1989) *The Informational City*. Oxford: Blackwell.

Castells, M. (1996) 'The Space of Flows', in *The Rise of the Network Society – The Information Age: Economy, Society, Culture, Vol. 1*. Oxford: Blackwell, 376–428.

CDMC (Central Manchester Development Corporation) (1989) 'Strategy for Consultation'. Manchester: Central Manchester Development Corporation.

CDMC (Central Manchester Development Corporation) (1996) '1988–1996: Eight Years of Achievement'. Manchester: Central Manchester Development Corporation.

Chakrabarty, D. (1992) 'Of Garbage, Modernity, and the Citizen's Gaze'. *Economic and Political Weekly* 27(10/11): 541–7.

Chase, J., M. Crawford, and J. Kaliski (1999) *Everyday Urbanism*. New York, NY: The Monacelli Press.

Chatterjee, I. (2009) 'Social Conflict and the Neoliberal City: A Case of Hindu–Muslim Violence in India'. *Transactions of the Institute of British Geographers* 34(2): 143–60.

Chion, M. (1999) *The Voice in Cinema*. New York, NY: Columbia University Press.

Chiping, H. (2013) 'Is HK to Become "Asia's Naples"'? *China Daily*, 11 July, available at: www.chinadailyasia.com/opinion/2013-07/11/content_15077577.html (accessed 4 September 2015).

Civic Trust (1964) *Derelict Land: A Study of Industrial Dereliction and How it May be Redeemed*. London: Civic Trust.

Clausen, L. (2003) *Global News Production*. Copenhagen: Copenhagen Business School Press.

Clemens, P. (2011) *Punching Out: One Year in a Closing Auto Plant*. New York, NY: Doubleday.

Coletto, D. (2010) *The Informal Economy and Employment in Brazil: Latin America, Modernization, and Social Changes*. London: Palgrave Macmillan.

Conley, T. (2007) *Cartographic Cinema*. Minneapolis, MN: University of Minnesota Press.

Conran Roche (1989) 'Pomona Docks Sub-Study for a CTC Facility', prepared for Central Manchester Development Corporation, London.

Cosgrove, D. (2001) *Apollo's Eye: A Cartographic Genealogy of the Earth in the Western Imagination*. Baltimore, MD: Johns Hopkins University Press.

Cosgrove, D. (2006) *Geographical Imagination and the Authority of Images*. Stuttgart: Franz Steiner Verlag.

Cottle, S. (2009) *Global Crisis Reporting: Journalism in the Global Age*. Maidenhead: Open University Press.

Coutinho, E. (dir.) (1992) *Boca de Lixo*.

Coward, M. (2008) *Urbicide: The Politics of Urban Destruction*. New York, NY: Routledge.

Crinson, M. (2005) 'Urban Memory: An Introduction', in *Urban Memory: History and Amnesia in the Modern City*. London and New York, NY: Routledge, xi–xxii.

Cronon, W. (1991) *Nature's Metropolis: Chicago and the Great West*. New York, NY and London: W.W. Norton and Co.

Cupers, K. and M. Miessen (2002) *Spaces of Uncertainty*. Wuppertal: Verlag Müller und Busmann.

D'Alisa, G. and M. Armiero (2013) 'What Happened to the Trash? Political Miracles and Real Statistics in an Emergency Regime'. *Capitalism Nature Socialism* 24(4): 29–45.

Dann, G.M.S. and A.V. Seaton (eds) (2002) *Slavery, Contested Heritage, and Thanatourism*. New York, NY: Routledge.

Davies, N. (2008) *Flat Earth News: An Award-Winning Reporter Exposes Falsehood, Distortion and Propaganda in the Global Media*. London: Vintage.

Davis, M. (1992) *City of Quartz*. New York, NY: Vintage.

Davis, M. (1998) *Ecology of Fear: Los Angeles and the Imagination of Disaster*. New York, NY: Vintage.

Davis, M. (2001) 'Sunshine and the Open Shop: Ford and Darwin in 1920s Los Angeles', in T. Sitton and W. Deverell (eds), *Metropolis in the Making: Los Angeles in the 1920*. Berkeley, CA: University of California Press, 96–122.

De Boeck, F. and M.F. Plissart (2004) *Kinshasa: Tales of the Invisible City*. Ludion and Tervuren: Royal Museum for Central Africa and Vlaams Architectuurinstitut Vai.

De Cauter, L. and M. Dehaene (eds) (2008) *Heterotopia and the City: Public Space in a Postcivil Society*. Abingdon: Routledge.

Deas, I., J. Peck, A. Tickell, K. Ward, and M. Bradford (1999) 'Rescripting Urban Regeneration, the Mancunian Way', in R. Imrie and H. Thomas (eds), *British Urban Policy: An Evaluation of the Urban Development Corporations*. London: SAGE Publications, 206–32.

Debord, G. (2004) *Society of the Spectacle*, trans. K. Knabb. London: Rebel Press.

Degen, M.M. (2008) *Sensing Cities: Regenerating Public Life in Barcelona and Manchester*, Routledge Studies in Human Geography. Abingdon and New York, NY: Routledge.

Del Rio, V. and D. de Alcantara (2009) 'The Cultural Corridor Project: Revitalization and Preservation in Downtown Rio de Janeiro', in V. del Rio and W.J. Siembieda (eds), *Contemporary Urbanism in Brazil: Beyond Brasilia*. Gainesville, FL: University Press of Florida.

DeLillo, D. (1999) *Underworld*. London: Picador.

Dennis, R. (2008) *Cities in Modernity: Representations and Productions of Metropolitan Space (1840–1930)*, Cambridge Studies in Historical Geography. Cambridge: Cambridge University Press.

Der Spiegel (1973) 'Impfung gegen Elend', 10 September, available at: www.spiegel.de/spiegel/print/d-41911250.html (accessed 7 September 2015).

Derraik, J.G. (2002) 'The Pollution of the Marine Environment by Plastic Debris: A Review'. *Marine Pollution Bulletin* 44(9): 842–52.

Derrida, J. (1988) *Limited Inc.*, trans. S. Weber. Chicago, IL: Northwestern University Press.

DeSilvey, C. and T. Edensor (2013) 'Reckoning With Ruins'. *Progress in Human Geography* 37(4): 1–21.

DETR (Department of the Environment, Transport and the Regions) (1999) *Towards an Urban Renaissance: Final Report of the Urban Task Force*. London: DETR.

Di Palma, V. (2014) *Wasteland: A History*. New Haven, CT and London: Yale University Press.

DiManno, R. (2011) 'That Trashy Odour in Naples? It's the Smell of Corruption'. *Toronto Star*, 6 May, available at: www.thestar.com/news/world/2011/05/06/that_trashy_odour_in_naples_its_the_smell_of_corruption.html (accessed 4 September 2015).

Dines, N. (2012a) *Tuff City: Urban Change and Contested Space in Central Naples*. New York, NY: Berghahn.

Dines, N. (2012b) 'Beyond the Aberrant City: Towards a Critical Ethnography of Naples'. *Lo Squaderno* 24: 21–5.

Dines, N. (2013) 'Bad News From an Aberrant City: A Critical Analysis of the British Press's Portrayal of Organised Crime and the Refuse Crisis in Naples'. *Modern Italy* 18(4): 409–22.

Doron, G.M. (2000) 'The Dead Zone and the Architecture of Transgression'. *City*, 4(2): 247–63.

Doron, G.M. (2007a) '…Badlands, Blank Space, Border Vacuum…'. *Field: A Free Journal of Architecture* 1(1): 10–23, available at: www.field-journal.org/uploads/file/2007_Volume_1/g%20doron.pdf (accessed 4 September 2015).

Doron, G.M. (2007b) 'Dead Zones, Outdoor Rooms and the Architecture of Transgression', in K.A. Franck and Q. Stevens (eds), *Loose Space: Possibility and Diversity in Urban Life*. Abingdon and New York, NY: Routledge, 210–29.

Douglas, M. (2002 [1966]) *Purity and Danger: An Analysis of the Concepts of Pollution and Taboo*. London and New York, NY: Routledge.

Dryzek, J. (2013) *The Politics of the Earth: Environmental Discourses*, 3rd ed. Oxford: Oxford University Press.

Edensor, T. (2005) *Industrial Ruins: Space, Aesthetics and Materiality*. Oxford and New York, NY: Berg.

Edensor, T. (2007) 'Social Practices, Sensual Excess and Aesthetic Transgression in Industrial Ruins', in K.A. Franck and Q. Stevens (eds), *Loose Space: Possibility and Diversity in Urban Life*. Abingdon and New York, NY: Routledge, 234–52.

Ehsani, K., A. Keshavarzian, and N.C. Moruzzi (2009) 'Tehran, June 2009'. *Middle East Research and Information Project*, 28 June, available at: www.merip.org/mero/mero062809 (accessed 4 September 2015).

El-Dahdah, F. (1998) 'On Solidere's Motto "Beirut: Ancient City of the Future"', in P. Rowe and H. Sarkis (eds), *Projecting Beirut: Episodes in the Construction and Reconstruction of a Modern City*. Munich: Prestel Verlag, 122–33.

Eleftheriou-Smith, L.M. (2014) 'Ebola Virus: Sierra Leone Boy's UK School Placement Cancelled over "Misguided Hysteria" by Parents over the Disease'. *Independent*, 8 October, available at: www.independent.co.uk/news/uk/home-news/sierra-leone-boys-uk-school-placement-cancelled-over-misguided-hysteria-by-parents-over-ebola-9781366.html (accessed 8 October 2014).

Elias, N. (1978) *The Civilising Process: The History of Manners: Sociogenetic and Psychogenetic Investigations*, trans. E. Jephcott. Oxford: Basil Blackwell.

El-Khoury, R. (1998) 'The Postwar Planning of Beirut', in P. Rowe and H. Sarkis (eds), *Projecting Beirut: Episodes in the Construction and Reconstruction of a Modern City*. Munich: Prestel Verlag, 183–6.

Elsaesser, T. (2009) '*Sonnen-Insulaner*: On a Berlin Island of Memory', in U. Staiger, H. Steiner, and A. Webber (eds), *Memory Culture and the Contemporary City*. New York, NY: Palgrave Macmillan.

Engels, F. (2009) *The Condition of the Working Class in England*, 3rd edn. London: Penguin Books.

Engler, M. (2004) *Designing America's Waste Landscapes*. Baltimore, MD: Johns Hopkins University Press.

English Tourism Board, LDR International (1989) 'Manchester, Salford and Trafford Strategic Development Initiative: A Framework for Tourism Development'. English Tourism Board, LDR International.

Entekhab.ir (2011) 'Sardar Radan: Polis Mitavanad Baraye Jamavari-ye Dish, Vorood be Manazel ham Dashte bashad [Sardar Radan: Police can also Enter the Houses in order to Collect the Satellite Dishes]', available at: www.entekhab.ir/fa/news/26303 (accessed 4 September 2015).

Environment Bureau (2013) *Hong Kong Blueprint for Sustainable Use of Resources 2013–2022*. Available at: www.enb.gov.hk/en/files/WastePlan-E.pdf (accessed 4 September 2015).

Epprecht, M. (1998) 'The "Unsaying" of Indigenous Homosexualities in Zimbabwe: Mapping a Blindspot in an African Masculinity'. *Journal of Southern African Studies* 24(4): 631–51.

Ethington, P. (2010) 'Ab Urbis Condita: Regional Regimes Since 13,000 Before Present', in W. Deverell and G. Hise (eds), *A Companion to Los Angeles*. Malden, MA: Blackwell, 177–215.

Evers, C. and K. Seale (eds) (2014) *Informal Urban Street Markets: International Perspectives*. New York, NY: Routledge.

ExUrbe (2013) 'Peel and the Liverpool City Region: Predatory Capitalism or Providential Corporatism?' Liverpool: ExUrbe.

Fanon, F. (1967 [1952]) *Black Skin, White Masks*. London: Pluto Press.

Fardon, R. (forthcoming) 'Purity as Danger: "Purity and Danger Revisited" Again!', in R. Duschinsky (ed.), *Purity and Impurity Across Anthropology, Psychology & Religious Studies: Contaminating Disciplines*. Basingstoke: Palgrave Macmillan.

Farley, P. and M. Symmons Roberts (2011) *Edgelands: Journeys into England's True Wildnerness*. London: Jonathan Cape.

Faucher, L. (1969) *Manchester in 1844: Its Present Condition and Future Prospects*, Cass Library of Industrial Classics. London: Frank Cass and Co., Ltd.

Field, A. (1913) *'Verb Sap': On Going to West Africa, to Northern, Southern Nigeria and to the Coasts*, 3rd edn. London: Bale, Sons and Danielsson.

Fisher, I. (2007) 'In Mire of Politics and the Mafia, Garbage Reigns', *New York Times*, 31 May, available at: www.nytimes.com/2007/05/31/world/europe/31naples.html?pagewanted=all (accessed 4 September 2015).

Fishman, R. (1977) *Urban Utopias in the Twentieth Century: Ebenezer Howard, Frank Lloyd Wright, and Le Corbusier*. New York, NY: Basic Books, Inc.

Flynn, H. (2013) 'Pomona Palace'. *Skyliner*, available at: www.theskyliner.org/pomona-palace/?rq=pomona%20palace (accessed 1 September 2015).

Flynn, H. (2014a) 'Pomona: The Lost Island of Manchester'. *Guardian*, 7 August, available at: www.theguardian.com/cities/2014/aug/07/pomona-lost-island-manchester-dockland-wasteland-oasis (accessed 1 September 2015).

Flynn, H. (2014b) 'Pomona Island on Film: Interview With George Haydock', *Skyliner*, available at: www.theskyliner.org/pomona-island-on-film/?rq=beauty%20of%20the%20desolate (accessed 1 September 2015).

Flynn, H. (2015) Personal interview with Brian Rosa.

Foley, M. and J. Lennon (eds) (2000) *Dark Tourism*. London: International Thomson Business Press.

Food and Environmental Hygiene Department (2012) *Controlling Officer's Environmental Report 2012*, available at: www.fehd.gov.hk/english/publications/environmental_report/2012/COER_2012_2.html (accessed 4 September 2015).

Food and Environmental Hygiene Department (2014) *Cleansing Services*, available at: www.fehd.gov.hk/english/pleasant_environment/cleansing/clean1.htm (accessed 4 September 2015).

Foucault, M. (1973) *The Order of Things: An Archaeology of the Human Sciences*. London: Vintage.

Foucault, M. (1977) *Discipline and Punish: The Birth of the Prison.* New York, NY: Vintage.

Foucault, M. (1998) 'Of Other Spaces', in J.D. Faubion (ed.), *Aesthetics, Method, and Epistemology.* New York, NY: The New Press.

Franck, K.A. and Q. Stevens (eds) (2006) *Loose Space: Possibility and Diversity in Urban Life.* London and New York, NY: Routledge.

Freeman, C. (2014) 'The Liberian Slum Where Ebola Spreads Death Among Killer Virus "Deniers"'. *Telegraph,* 8 August, available at: www.telegraph.co.uk/news/worldnews/africaandindianocean/liberia/11020768/The-Liberian-slum-where-Ebola-spreads-death-among-killer-virus-deniers.html (accessed 16 October 2014).

Fregonese, S. (2009) 'The Urbicide of Beirut? Geopolitics and the Built Environment in the Lebanese Civil War (1975–1976)'. *Political Geography* 28(5): 309–18.

Freitas, M. (2013) 'Fim da Perimetral não extingue tradição da Feira da Praça XV', available at: http://puc-riodigital.com.puc-rio.br/Jornal/Cidade/Fim-da-Perimetral-nao-extingue-tradicao-da-Feira-da-Praca-XV-23698.html#.U_QX1PbXYl8 (accessed 4 September 2015).

Frickel, S. (2012) 'Missing New Orleans: Tracking Knowledge and Ignorance Through the Urban Hazardscape', in S. Foote and E. Mazzolini (eds), *Histories of the Dustheap: Waste Material Cultures, Social Justice.* Cambridge, MA: MIT Press.

Fukuyama, F. (1989) 'The End of History'. *National Interest* (Summer 1989), available at: https://ps321.community.uaf.edu/files/2012/10/Fukuyama-End-of-history-article.pdf (accessed 7 September 2015).

Furtado, J. (dir.) (1989) *Ilha das Flores.* Brazil.

Gabbe, B. (2007) 'Cholera-Angst in Neapel'. *Hamburger Abendblatt,* 24 May, available at: www.abendblatt.de/vermischtes/article860843/Cholera-Angst-in-Neapel.html (accessed 4 September 2015).

Gallagher, J. (2013) *Revolution Detroit: Strategies for Urban Reinvention.* Detroit, MI: Wayne State University Press.

Galtung, J. and M.H. Ruge (1965) 'The Structure of Foreign News: The Presentation of the Congo, Cuba and Cyprus Crises in Four Norwegian Newspapers'. *Journal of Peace Research* 2(1): 64–91.

Gandy, M. (2013) 'Marginalia: Aesthetics, Ecology, and Urban Wastelands'. *Annals of the Association of American Geographers* 103(6): 1301–16.

Garrett, B. (2013) *Explore Everything: Place-Hacking the City.* London: Verso.

Gaviria, V. (dir.) (1998) *La Vendedora de Rosas.*

Genuske, A. (2012) 'Trashcam Project Uses Dumpsters As Cameras'. *The Huffington Post,* 21 April, available at: www.huffingtonpost.com/2012/04/21/trashcam-project_n_1441416.html (accessed 4 September 2015).

George, R. (2008) *The Big Necessity: The Unmentionable World of Human Waste and Why it Matters.* New York, NY: Holt.

Gidwani, V. and R.N. Reddy (2011) 'The Afterlives of "Waste": Notes from India for a Minor History of Capitalist Surplus'. *Antipode* 43(5): 1625–58.

Gilbert, A. (2007) 'The Return of the Slum: Does Language Matter?' *International Journal of Urban and Regional Research* 31(4): 697–713.

Giles, D.B. (2011) 'Dumpsters, Forbidden Fruit, and the Social Afterlife of Things'. University of Washington Department of Anthropology E-News, available at: http://depts.washington.edu/anthweb/enews/2011-05/dumpsters.php (accessed 4 September 2015).

Giles, D.B. (2014) 'The Anatomy of a Dumpster: Abject Capital and the Looking Glass of Value'. *Social Text* 32(1 118), 93–113.

Gille, Z. (2007) *From the Cult of Waste to the Trash Heap of History: The Politics of Waste in Socialist and Postsocialist Hungary.* Bloomington, IN: Indiana University Press.

Godfrey, B.J. (2013) 'Urban Renewal, Favelas, and Guanabara Bay: Environmental Justice and Sustainability in Rio de Janeiro', in I. Vojnovic (ed.), *Urban Sustainability: A Global Urban Context*, East Lansing, MI: Michigan State University Press, 359–86.

Goldstein, J. (2013) 'Terra Economica: Waste and the Production of Enclosed Nature'. *Antipode* 45(2): 357–75.

Google Maps (2013) 'Take a Stroll Through Abandoned "Battleship Island" on Google Maps', available at: http://google-latlong.blogspot.ca/2013/06/take-stroll-through-abandoned.html (accessed 4 September 2015).

Graham, S. (2003) 'Lessons in Urbicide'. *New Left Review* 19, 63–77.

Graham, S. (2004) 'Cities as Strategic Sites: Place Annihilation and Urban Geopolitics', in S. Graham (ed.), *Cities, War and Terrorism: Towards an Urban Geopolitics.* Malden, MA and Oxford: Blackwell, 31–53.

Graham, S. and S. Marvin (2001) *Splintering Urbanism: Networked Infrastructures, Technological Mobilities and the Urban Condition.* London and New York, NY: Routledge.

Gray, A. (2004) *Unsocial Europe: Social Protection or Flexploitation?* Ann Arbor, MI: Pluto Press.

Greenfield, L. (dir.) (2012) *The Queen of Versailles.*

Gregson, N. and L. Crewe (2003) *Second-Hand Cultures.* London: Bloomsbury.

Gribaudi, G. (2008) 'Il Ciclo Vizioso dei Rifiuti Campani'. *Il Mulino* 53(1): 17–33.

Groth, J. and E. Corijn (2005) 'Reclaiming Urbanity: Indeterminate Spaces, Informal Actors and Urban Agenda Setting'. *Urban Studies* 42(3): 503–26.

Guardian (2010) 'Photographer Pietro Masturzo's Best Shot: Interview by Andrew Pulver', 8 December, available at: www.theguardian.com/artanddesign/2010/dec/08/photography-pietro-masturzo-best-shot (accessed 24 February 2013).

Habermas, J. (1989) *The Structural Transformation of the Public Sphere: An Inquiry into a Category of Bourgeois Society.* Cambridge, MA: MIT Press.

Hadjithomas, J. and K. Joreige (dirs) (2008) *Je Veux Voir*, 75.

Hall, M. (2013) 'The Ecological and Environmental Significance of Urban Wastelands and Drosscapes', in M.J. Zapata and M. Hall (eds), *Organising Waste in the City.* Bristol: Policy Press, 21–41.

Hall, P. (1998) *Cities in Civilization: Culture, Innovation and Urban Order.* London: Phoenix Giant.

Hall, S. (1994) 'Cultural Identity and Diaspora', in P. Williams and L. Chrisman (eds), *Colonial Discourse and Postcolonial Theory: A Reader.* Harlow: Harvester Wheatsheaf, 392–403.

Hall, S., C. Critcher, T. Jefferson, J. Clarke, and B. Roberts (1978) *Policing the Crisis: Mugging, the State and Law and Order.* London: Macmillan.

Hamon, P. (1992) *Expositions: Literature and Architecture in Nineteenth-Century France*, trans. K. Sainson-Frank and L. Maguire. Berkeley, CA, Los Angeles, CA and Oxford: University of California Press.

Hannerz, U. (2004) *Foreign News: Exploring the World of Foreign Correspondents.* Chicago, IL: University of Chicago Press.

Harcup, T. and D. O'Neill (2001) 'What is News? Galtung and Ruge Revisited'. *Journalism Studies* 2(2): 261–80.

Harkness, M. (1889) *Captain Lobe: A Story of the Salvation Army.* London: Hodder and Stoughton.

Harper, T. (2013) 'The Biggest Company You've Never Heard of: Lifting the Lid on Peel Group – the Property Firm Owned by Reclusive Tax Exile John Whittaker'. *Independent*. 18 October, available at: www.independent.co.uk/news/uk/home-news/the-biggest-company-youve-never-heard-of-lifting-the-lid-on-peel-group--the-property-firm-owned-by-reclusive-tax-exile-john-whittaker-8890201.html (accessed 7 September 2015).

Harvey, D. (1982) *The Limits to Capital*. Oxford: Blackwell.

Harvey, D. (1989a) 'From Managerialism to Entrepreneurialism: The Transformation in Urban Governance in Late Capitalism'. *Geografiska Annaler B* 71(1): 3–17.

Harvey, D. (1989b) *The Condition of Postmodernity: An Enquiry into the Origins of Cultural Change*. Oxford: Blackwell.

Harvey, D. (2006) *Spaces of Global Capitalism: A Theory of Uneven Geographical Development*. London and New York, NY: Verso.

Haugbolle, S. (2010) *War and Memory in Lebanon*. Cambridge: Cambridge University Press.

Hawkins, G. (2006) *The Ethics of Waste: How We Relate to Rubbish*. Lanham, MD: Rowman & Littlefield.

Hawkins, G. (2007) 'Waste in Sydney: Unwelcome Returns'. *PMLA* 122(1): 348–51.

Hawkins, G. and S. Muecke (2003) *Culture and Waste: The Creation and Destruction of Value*. Lanham, MD: Rowman & Littlefield.

Hebbert, M. (2010) 'Manchester: Making it Happen', in J. Punter (ed.), *Urban Design and the British Urban Renaissance*. London and New York, NY: Routledge, 51–67.

Heddaya, M. (2014) 'US Senator Proposes Conservation Expanse for Michael Heizer's Land Art'. *Hyperallergic*, 4 November, available at: http://hyperallergic.com/160604/us-senator-proposes-conservation-expanse-for-michael-heizers-land-art/ (accessed 7 September 2015).

Heizer, M. (2014) Personal correspondence.

Herman, E.S. and R.W. McChesney (1997) *The Global Media: The New Missionaries of Corporate Capitalism*. London: Cassell.

Hiebert, D., J. Rath, and S. Vertovec (2014) 'Urban Markets and Diversity: Towards a Research Agenda'. *Ethnic and Racial Studies* 38(1): 5–21.

High, S. and D.W. Lewis (2011) *Corporate Wasteland: The Landscape and Memory of Deindustrialization*. Ithaca, NY: Cornell University Press.

Highmore, B. (2002) *Everyday Life and Cultural Theory: An Introduction*. London: Routledge.

Hird, M.J. (2012) 'Knowing Waste: Towards an Inhuman Epistemology'. *Social Epistemology* 26(3/4), 453–69.

HM Stationery Office (1963) 'New Life for Dead Lands'. London: Ministry of Housing and Local Government.

Hodding Carter, W. (2006) *Flushed: How the Plumber Saved Civilization*. New York, NY: Atria Books.

Hofmann, M., J.R. Westermann, I. Kowarik, and E. van der Meer (2012) 'Perceptions of Parks and Urban Derelict Land by Landscape Planners and Residents'. *Urban Forestry & Urban Greening* 11(3): 303–12.

Hogen-Esch, T. (2010) 'Consolidation, Fragmentation, and the New Fiscal Federalism', in W. Deverell and G. Hise (eds), *A Companion to Los Angeles*. Malden, MA: Blackwell, 233–49.

Hohn, D. (2011) *Moby-Duck*. New York, NY: Penguin.

Hommels, A. (2005) *Unbuilding Cities: Obduracy in Socio-Technical Change*. Cambridge, MA: MIT Press.

Hooper, J. (2008) 'More Garbage From Naples'. *Guardian*, 10 January, available at: www.theguardian.com/commentisfree/2008/jan/10/moregarbagefromnaples#history-link-box (accessed 4 September 2015).

Hoornweg, D., P. Bhada-Tata, and C. Kennedy (2013) 'Environment: Waste Production Must Peak this Century'. *Nature* 502: 615–17.

Hope-Nicholson, M. (1959) *Mountain Gloom and Mountain Glory: The Development of the Aesthetics of the Infinite*. New York, NY: Cornell University Press.

Hubbard, P. (2006) *City*. London: Routledge.

Hulme Lever, W. (1921) 'Letters from William Hulme Lever to Lord Leverhulme: Nigerian tour 1921: Letter No. 5, 5 August 1921'. United Africa Company Leverhulme Business Correspondence, Box 1376 TT 3810, Location SR1 Lever, WH.

Humes, E. (2012) *Garbology: Our Dirty Love Affair with Trash*. New York, NY: Avery.

Hunt, J. (2012) 'Vik Muniz: Pictures of Junk, Jardim Gramacho, Rio de Janeiro'. *Art and Architecture Journal*, 24 February, available at: https://aajpress.wordpress.com/2012/02/24/vik-muniz-pictures-of-junk-jardim-gramacho-rio-de-janeiro/ (accessed 4 September 2015).

Hurley, A. (1995) *Environmental Inequalities: Class, Race and Industrial Pollution in Gary, Indiana, 1945–1980*. Chapel Hill, NC: University of North Carolina Press.

Huyssen, A. (1994) *Twilight Memories: Marking Time in a Culture of Amnesia*. New York, NY: Routledge.

Huyssen, A. (2003) *Present Pasts: Urban Palimpsests and the Politics of Memory*. Stanford, CA: Stanford University Press.

Huyssen, A. (2008) 'Introduction: World Cultures, World Cities', in A. Huyssen (ed.) *Other Cities, Other Worlds*. Durham, NC and London: Duke University Press.

Huyssen, A. (2010) 'Authentic Ruins: Products of Modernity', in J. Hell and A. Schönle (eds), *Ruins of Modernity*. Durham, NC and London: Duke University Press.

Ibarz, V. (2009) 'Entrevisa a Daniel Canogar: Otras geologías'. *Disturbis*, available at: www.disturbis.esteticauab.org/Disturbis567/VIbarz.html (accessed 4 September 2015).

Jaber, R. (2006) *Bīrītūs: madīnah taḥt al-'arḍ* [Berytus: Underground City]. Beirut: Dar al-Adab.

Jiji Press (2014) 'South Korea Wants Heritage Status Denied'. *The Japan Times*, 5 February, available at: www.japantimes.co.jp/news/2014/02/05/national/south-korea-wants-heritage-status-denied/#.Ve2HwPSTXgo (accessed 7 September 2015).

Johnson, B. (2013) *Zero Waste Home: The Ultimate Guide to Simplifying Your Life by Reducing Your Waste*. New York, NY: Scribner.

Johnson, B. (2014) 'What's in Our Family's Jar of Annual Waste?' *Zero Waste Home*, available at: www.zerowastehome.com/2014/11/whats-in-our-familys-jar-of-annual-waste.html (accessed 25 November 2014).

Jordan, C. (2005) 'Intolerable Beauty: Portraits of American Mass Consumption (2003–2005) – About'. *Chris Jordan, Photographic Arts*, available at: http://chrisjordan.com/gallery/intolerable/#about (accessed 4 September 2015).

Jorgensen, A. (2008) *Urban Wildscapes*. Department of Landscape, Sheffield, UK: University of Sheffield & Environment Room Ltd.

Jorgensen, A. (2012) 'Introduction', in A. Jorgensen and R. Keenan (eds), *Urban Wildscapes*. Abingdon and New York, NY: Routledge, 1–14.

Jorgensen, A. and M. Tylecote (2007) 'Ambivalent Landscapes: Wilderness in the Urban Interstices'. *Landscape Research* 32(4): 443–62.

Joyce, P. (2003) *The Rule of Freedom: Liberalism and the Modern City*. London and New York, NY: Verso.

Kam, L. (2010) 'Hong Kong on the Move: Creating Global Cultures', in L. Kam (ed.), *Hong Kong Culture: Word and Image*. Hong Kong: Hong Kong University Press, 1–7.

Kantaris, G. (2003) 'The Young and the Damned: Street Visions in Latin American Cinema', in S.M. Hart and R. Young (eds), *Contemporary Latin American Cultural Studies*. London: Arnold, 177–89.

Kappelin, K. (2008) 'Maffian tjänar på Neapels miljökaos', *SVT Nyheter*, 9 January, available at: www.svt.se/nyheter/maffian-tjanar-pa-neapels-miljokaos (accessed 4 September 2015).

Kaye, L. (2011) 'Community Groups Tackle Naples' Rubbish Problem'. *Guardian*, 18 August, available at www.theguardian.com/sustainable-business/naples-waste-management-problem-community-groups (accessed 4 September 2015).

Kaye, L. (2014) 'Spring Cancelled: Trees Crushed at Pomona'. Manchester: Pirate Party UK.

Keeling, N. (2014) 'Manchester's "Eden Project": Protest picnic to protect Pomona docks'. *Manchester Evening News*, 18 March, available at: www.manchestereveningnews.co.uk/news/greater-manchester-news/manchesters-eden-project-protest-picnic-6846772 (accessed 7 September 2015).

Kennedy, C. (2007) 'Urban Metabolism'. *The Encyclopedia of Earth*, available at: www.eoearth.org/view/article/51cbef227896bb431f69c9a5 (accessed 14 April 2015).

Kennedy, G. (2007) *An Ontology of Trash: The Disposable and Its Problematic Nature*. New York, NY: State University of New York Press.

Kerbaj, M. (2013) *Lettre à la Mère*. Comic book. Bresson: l'Apocalypse.

Kilpatrick, H. (2000) 'Literary Creativity and the Cultural Heritage: The *Aṭlāl* in Modern Arabic Fiction', in K. Abdel-Malek and W. Hallaq (eds), *Tradition, Modernity, and Postmodernity in Arabic Literature: Essays in Honor of Professor Issa J. Boullata*. Leiden, Boston, MA and Cologne: Brill, 28–44.

Kimmelman, M. (1999) 'A Sculptor's Colossus of the Desert'. *New York Times*, 12 December, available at: www.nytimes.com/library/arts/121299heizer-art.html (accessed 7 September 2015).

Kimmelman, M. (2005) 'Art's Last, Lonely Cowboy'. *New York Times*, 6 February, available at: www.nytimes.com/2005/02/06/magazine/arts-last-lonely-cowboy.html?_r=0 (accessed 7 September 2015).

Kingsbury, P. and J.P. Jones (2009) 'Walter Benjamin's Dionysian Adventures on Google Earth'. *Geoforum* 40: 502–13.

Kitchen, T. (1997) *People, Politics, Policies and Plans: The City Planning Process in Contemporary Britain*. London: Paul Chapman Publishing.

Knight, C. (2009) 'Michael Heizer's "City" May Not Glow in the Dark'. *Los Angeles Times* – Culture Monster, 28 February, available at: http://latimesblogs.latimes.com/culturemonster/2009/02/heizer-yucca.html (accessed 7 September 2015).

Knox, T.M. (1924–1927) 'Niger Company Ltd.: Diary of Tour Through the Congo and West Africa'. United Africa Company (Unilever Archives and Records), 2/34/4/1/1.

Knox, T.M. (1929) 'Niger Company Ltd: Reports of T.M. Knox to Chairman and Board on His Tour Through West Africa'. United Africa Company (Unilever Archives and Records), 2/34/4/1/3.

Koolhaas, R. (2002) 'Junkspace'. *October* 100(Spring): 175–90.

Kristeva, J. (1982) *Powers of Horror: An Essay on Abjection*. New York, NY: Columbia University Press.

Kubler, G. (1962) *The Shape of Time: Remarks on the History of Things*. New Haven, CT: Yale University Press.

Kuebler-Wolf, E. (2013) 'Land Art'. Unpublished lecture, University of Saint Francis, USA.

Kurgan, L. (2013) *Close Up at a Distance: Mapping, Technology, and Politics*. Brooklyn, NY: Zone Books.

Lafargue, P. (1907) *The Right to Be Lazy, and Other Studies*. Chicago, IL: Charles H. Kerr & Company.

Lafargue, P. (1969 [1880]) *Le Droit à la Paresse*. Paris: F. Maspero.

Lampugnani, V.M., E.G. Pryor, S.H. Pau, T. Spengler, and P. Zachmann (1993) *Hong Kong Architecture: The Aesthetics of Density*. Munich: Prestel Pub.

Landau, E., Z. Verjee, and A. Mortensen (2014) 'Uganda President: Homosexuals are "Disgusting"'. *CNN*, 25 February, available at: http://edition.cnn.com/2014/02/24/world/africa/uganda-homosexuality-interview/ (accessed 15 October 2014).

Laporte, D. (2000) *History of Shit*. Cambridge, MA: MIT Press.

Lash, S. and J. Urry (1987) *The End of Organized Capitalism*. Cambridge: Polity.

Lash, S. and J. Urry (1994) *Economies of Signs and Space*. Thousand Oaks: Sage.

Le Corbusier (1967 [1933]) *Radiant City*. New York, NY: Orion Press.

Leary, M.E. (2008) 'Gin and Tonic or Oil and Water: The Entrepreneurial City and Sustainable Managerial Regeneration in Manchester'. *Local Economy* 23(3): 222–33.

Leary, M.E. (2011) *The Production of Urban Public Space: A Lefebvrian Analysis of Castlefield, Manchester*. London: Department of Sociology, Goldsmiths, University of London.

LeDuff, C. (2013) *Detroit: An American Autopsy*. London: Penguin Press.

Lee, K.M. and H. Wong (2004) 'Marginalized Workers in Postindustrial Hong Kong'. *The Journal of Comparative Asian Development* 3(2), 249–80.

Lefebvre, H. (1991) *Critique of Everyday Life: Volume 1*. London: Verso.

Lefebvre, H. (1996 [1968]) *Writings on Cities*, ed. and trans. E. Kofman and E. Lebas. Malden, MA: Blackwell.

Lefebvre, H. (2009 [1974]) *The Production of Space*, trans. D. Nicholson-Smith. Oxford: Blackwell.

Leibig, J. von (1843) 'Leibig's Letters on Chemistry'. *The Spectator*, 14 October, 18–19, available at: http://archive.spectator.co.uk/article/14th-october-1843/18/liebig-s-letters-on-chemistry (accessed 7 September 2015).

Leone, M. (2012) 'My Schoolmate: Protest Music in Present-day Iran'. *Critical Discourse Studies* 9(4): 347–62.

Leung, C.Y. (2008) 'Everyday Life Resistance in a Post-Colonial Global City – A Study of Two Illegal Hawker Agglomerations in Hong Kong'. Unpublished doctoral dissertation, Hong Kong University of Science and Technology.

Levine, P. and A. Moore (2010) *Detroit Disassembled*. Akron, OH: Damiani/Akron Art Museum.

Lewis-Coleman, D. (2008) *Race Against Liberalism: Black Workers and the UAW in Detroit*. Urbana, IL: University of Illinois Press.

Lindner, C. and B. Rosa (eds) (forthcoming) *Deconstructing the High Line: Essays on Postindustrial Urbanism*. New Brunswick, NJ: Rutgers University Press.

Linsky, B. (1953) 'Air Pollution and Man's Health in Detroit'. *Public Health Report* 68(9): 870–2.

Locatelli, G. (2008) 'Immondizia in Prima Pagina 'è una Vergogna per l'Italia'. *La Repubblica (Naples edition)*, 10 January, available at: http://ricerca.repubblica.it/repubblica/archivio/repubblica/2008/01/10/immondizia-in-prima-pagina-una-vergogna.html?ref=search (accessed 4 September 2015).

Locke, J. (1988) *Two Treatises of Government*. Cambridge: Cambridge University Press.

Loughran, K. (2014) 'Parks for Profit: The High Line, Growth Machines, and the Uneven Development of Urban Public Spaces'. *City & Community* 13(1): 49–68.

Loures, L. (2015) 'Post-Industrial Landscapes as Drivers for Urban Redevelopment: Public Versus Expert Perspectives Towards the Benefits and Barriers of the Reuse of Post-industrial Sites in Urban Areas'. *Habitat International* 45(2): 72–81.

Lovejoy, L. (1912) 'Garbage and Rubbish'. *Proceedings of the Academy of Political Science in the City of New York* 2(3), National Housing Association, 62–9.

Lynch, K. (1960) *The Image of the City*. Cambridge, MA and London: MIT Press.

Lynch, K. (1972) *What Time is this Place?* Cambridge, MA and London: MIT Press.

Lynch, K. (1990) *Wasting Away.* San Francisco, CA: Sierra Club Books.

Lyotard, J.F. (1991) *The Inhuman.* London: Polity.

Ma, E. (2008) 'Transborder Visuality: The Changing Patterns of Visual Exchange Between Hong Kong and South China', in H.F. Siu and A.S. Ku (eds) *Hong Kong Mobile: Making a Global Population.* Hong Kong: Hong Kong University Press, 63–82.

Mabey, R. (1973) *The Unofficial Countryside*. London: Collins.

MacKay, J. (2014) Private correspondence.

Madanipour, A. (1998) *Tehran: The Making of a Metropolis.* Chichester: John Wiley & Sons.

Madgin, R. (2010) 'Reconceptualising the Historic Urban Environment: Conservation and Regeneration in Castlefield, Manchester 1960–2009'. *Planning Perspectives* 25(1): 29–48.

Makdisi, S. (1997) 'Laying Claim to Beirut: Urban Narrative and Spatial Identity in the Age of Solidere'. *Critical Inquiry* 23(3), Front Lines/Border Posts (Spring 1997): 660–705.

Manchester Evening News (2004) 'The Start of the Stop', 13 August, available at: www.manchestereveningnews.co.uk/news/greater-manchester-news/the-start-of-the-stop-1102002 (accessed 3 September 2015).

Manoukian, S. (2010) 'Where Is This Place? Crowds, Audio-vision, and Poetry in Post-election Iran'. *Public Culture* 22(2): 237–63.

Maranga-Musonye, M. (2014) 'Literary Insurgence in the Kenyan Urban Space: *Mchongoano* and the Popular Art Scene in Nairobi', in S. Newell and O. Okome (eds), *Popular Culture in Africa: The Episteme of the Everyday.* Abingdon and New York, NY: Routledge, 195–218.

Marchal, J. (2008) *Lord Leverhulme's Ghosts: Colonial Exploitation in the Congo*, London and New York, NY: Verso.

Marchand, Y. and R. Meffre (2010) *The Ruins of Detroit.* Göttingen: Steidl Verlag.

Marx, K. (1975) *The Eighteenth Brumaire of Louis Bonaparte: With Explanatory Notes.* New York, NY: International Publishers.

Marx, K. (1976) *Capital: A Critique of Political Economy*. London and New York, NY: Penguin Books in association with New Left Review.

Marx, K. (1984) *The Eighteenth Brumaire of Louis Bonaparte*. London: Lawrence and Wishart.

Marx, K. (1990) *Capital: A Critique of Political Economy*. London and New York, NY: Penguin Books, in association with New Left Review.

Marx, K. (1993) *Grundrisse: Foundations of the Critique of Political Economy (Rough Draft)*, trans. M. Nicolaus. Harmondsworth: Penguin.

Massey, D. (2005) *For Space*. London: SAGE Publications.

McCarthy, J. (2013) 'We have never been "post-political"'. *Capitalism Nature Socialism* 24(1): 19–25.

McClear, S. (2013) 'Punk's Influence on Fashion is on Display at Metropolitan Museum of Art's "Punk: Chaos to Couture"'. *Daily News*, 5 May, available at: www.nydailynews.com/life-style/fashion/punk-fashions-display-met-punk-chaos-couture-article-1.1334338 (accessed 25 September 2014).

McClintock, A. (1995) *Imperial Leather: Race, Gender and Sexuality in the Colonial Contest.* London and New York, NY: Routledge.

McCurry, J. (2008) 'Climate Change: How the Quest for Zero Waste Community Means Sorting the Rubbish 34 Ways'. *Guardian*, 5 August, available at: www.theguardian.com/environment/2008/aug/05/recycling.japan (accessed 4 September 2015).

McDonogh, G. and C. Wong (2005) *Global Hong Kong.* New York, NY and Abingdon: Routledge.

Meirelles, F. and K. Lund (dir.) (2002) *Cidade de Deus.*

Melosi, M. (2001) *Effluent America: Cities, Industry, Energy, and the Environment.* Pittsburgh, PA: University of Pittsburgh Press.

Melosi, M. (2002) 'The Fresno Sanitary Landfill in an American Cultural Context'. *The Public Historian* 24(3): 17–35.

Melosi, M. (2005) *Garbage in the Cities: Refuse, Reform and the Environment,* rev. edn. Pittsburgh, PA: University of Pittsburgh Press.

Mendes, S. (dir.) (2012) *Skyfall.*

Merrifield, A. (1993) 'The Struggle Over Place: Redeveloping American Can in Southeast Baltimore'. *Transactions of the Institute of British Geographers* 18(1): 102–21.

Miller, S.B. (2004) *Disgust: The Gatekeeper Emotion.* Hillsdale, NJ: Analytic Press.

Miller, V. (2006) 'The Unmappable: Vagueness and Spatial Experience'. *Space and Culture* 9(4): 453–67.

Miller, W.I. (1997) *The Anatomy of Disgust.* Cambridge, MA and London: Harvard University Press.

Millington, N. (2013) 'Post-Industrial Imaginaries: Nature, Representation and Ruin in Detroit, Michigan: Nature, Representation and Ruin in Detroit'. *International Journal of Urban and Regional Research* 37(1): 279–96.

Mills, C. (2001) 'Black Trash', in L. Westra and B. Lawson (eds), *Faces of Environmental Racism: Confronting Issues of Global Justice.* New York, NY: Rowman & Littlefield, 73–91.

Minter, A. (2013) *Junkyard Planet: Travels in the Billion-Dollar Trash Trade.* New York, NY: Bloomsbury.

Minton, A. (2009) *Ground Control: Fear and Happiness in the Twenty-first Century City.* London: Penguin Books.

Mirzoeff, N. (ed.) (2002) *The Visual Culture Reader.* London: Routledge.

Mitchell, T. (1994) *Picture Theory: Essays on Verbal and Visual Representation.* Chicago, IL: University of Chicago Press.

Moore, S.A. (2012) 'Garbage Matters: Concepts in New Geographies of Waste'. *Progress in Human Geography* 36(6): 780–99.

Mörtenböck, P. and H. Mooshammer (2008) 'Spaces of Encounter: Informal Markets in Europe'. *Architectural Research Quarterly* 12(3/4): 347–57.

Morton, T. (2010) 'Queer Ecology'. *PMLA* 125(2): 273–82.

Mroué, R. and F. Toufic (2007) *How Nancy Wished That Everything Was an April Fool's Joke.* Play. Seen at Home Works IV Festival, 15April 2008.

Mumford, L. (1961) *The City in History.* London: Penguin Books.

Muniz, V. and L. Walker (2010) *Waste Land (Press Notes)*, available at: www.wasteland-movie.com/downloads.html (accessed 19 October 2014).

Musella, A. (2008) *Mi Rfiuto! Le Lotte in Difesa della Salute e dell'Ambiente in Campania*. Rome: Sensibili alle Foglie.

Nabarro, R. (1980) 'The General Problem of Urban Wasteland'. *Built Environment* 6(3): 159–65.

Naccache, A.F.H. (1998) 'Beirut's Memorycide', in L. Meskell (ed.), *Archaeology Under Fire: Nationalism, Politics and Heritage in the Eastern Mediterranean and Middle East*. London: Routledge, 140–58.

Nagle, R. (2013) *Picking Up: On the Streets and Behind the Trucks With the Sanitation Workers of New York City*. New York, NY: Farrar, Straus & Giroux.

Newell, S. (2006) *The Forger's Tale: The Search for Odeziaku*. Athens, OH: Ohio University Press.

No Impact Man (2010) New York, NY: Oscilloscope Laboratories.

Noys, B. (2005) *The Culture of Death*. Oxford and New York, NY: Berg.

O'Brien, M. (2011) *A Crisis of Waste? Understanding the Rubbish Society*. New York, NY: Routledge.

Orgad, S. (2012) *Media Representation and the Global Imagination*. Cambridge: Polity.

Owen, M. (1938) 'Developments in Sewage Sludge Incineration'. *Sewage Works Journal* 10(1): 100–5.

Paes Henriques, R. (2008) 'A Psicose em Cena: Schreber, Aimée, Joyce e Estamira', in *Web-Based publications of the 9th Brazilian Congress of Fundamental Psychopathology*, 2008. IX Congresso Brasileiro de Psicopatologia Fundamental. Universidade Federal Fluminense: Psicopatologia Fundamental (website), available at: www.psico-patologiafundamental.org/pagina-temas-livres-459 (accessed 24 October 2014).

Palmer, A. and H. Clark (eds) (2005) *Old Clothes, New Looks: Second Hand Fashion*. Oxford and New York, NY: Berg.

Parkinson-Bailey, J.J. (2000) *Manchester: An Architectural History*. Manchester and New York, NY: Manchester University Press.

Parks, L. (2009) 'Digging Into Google Earth: An Analysis of "Crisis in Darfur"'. *Geoforum* 40: 535–45.

Parks, L. (2013) 'Mapping Orbit: Toward a Vertical Public Space', in C. Berry, J, Harbord, and R.O. Moore (eds), *Public Space, Media Space*. New York, NY: Palgrave.

Pasotti, E. (2010) 'Sorting through the Trash: The Waste Management Crisis in Southern Italy'. *South European Society and Politics* 15(2): 289–307.

Peck, J. and K. Ward (2002) 'Placing Manchester', in *City of Revolution: Restructuring Manchester*. Manchester and New York, NY: Manchester University Press, 1–17.

Pellow, D. (2004) *Garbage Wars: The Struggle For Environmental Justice in Chicago*. Cambridge, MA: MIT Press.

Pellow, D. and J. Brulle (2005) *Power, Justice and the Environment: A critical Appraisal of the Environmental Justice Movement*. Cambridge, MA: MIT Press.

Petrillo, A. (ed.) (2009) *Biopolitica di un Rifiuto. Le Rivolte Anti-discarica a Napoli e in Campania*. Verona: Ombre Corte.

Picon, A. (2000) 'Anxious Landscapes: From the Ruin to Rust'. *Grey Room* 1: 64–83.

Pine, J.B. and J.H. Gilmore (eds) (1999) *The Experience Economy: Work is Theatre & Every Business a Stage*. New York, NY: Harvard Business School Press.

Piqué, E. (2008) 'Nápoles, Entre la Basura e la Intolerancia Social'. *La Nación*, 21 May, 4, available at: http://servicios.lanacion.com.ar/archivo/2008/05/21/cuerpo-principal/004 (accessed 4 September 2015).

Pivaro, N. (2008a) 'My Day Trip to the Stunning Red Sands of Ordsall'. *Salford Star*, 8: 32–35, available at: www.salfordstar.com/media/ss8.pdf (accessed 1 September 2015).

Pivaro, N. (2008b) 'Costa del Salford'. *Salford Star*, 8: 31, available at: www.salfordstar.com/media/ss8.pdf (accessed 1 September 2015).

Platt, H.L. (2005) *Shock Cities: The Environmental Transformation and Reform of Manchester and Chicago.* Chicago, IL: The University of Chicago Press.

Polanyi, K. (2007) *The Great Transformation: The Political and Economic Origins of our Time*, rev. ed. Boston, MA: Beacon Press.

Ponting, C. (1991) *A New Green History of the World: The Environment and the Collapse of Great Civilization.* New York, NY: Penguin.

Popham, P. (2008) 'The Rotten Heart of Italy: See Naples and Die (of the Stench)'. *Independent*, 11 January, available at: www.independent.co.uk/news/world/europe/the-rotten-heart-of-italy-see-naples-and-die-of-the-stench-769448.html# (accessed 4 September 2015).

Prado, M. (dir.) (2004) *Estamira*.

Pratt, M.L. (1992) *Imperial Eyes: Travel Writing and Transculturation.* London and New York, NY: Routledge.

Prunier, G. (1998) *The Rwanda Crisis: History of a Genocide*, 2nd rev. edn. London: Hurst.

Puar, J. (ed.) (2012) 'Precarity Talk: A Virtual Roundtable with Lauren Berlant, Judith Butler, Bojana Cvejić, Isabell Lorey, Jasbir Puar, and Ana Vujanović'. *The Drama Review* 56(4), 163–77.

Quilley, S. (1999) 'Entrepreneurial Manchester: The Genesis of Elite Consensus'. *Antipode* 31(2): 185–211.

Quilley, S. (2002) 'Entrepreneurial Turns: Municipal Socialism and After', in J. Peck and K. Ward (eds), *City of Revolution: Restructuring Manchester.* Manchester and New York, NY: Manchester University Press, 76–94.

Rabbath, G.H. (2009) *Can One Man Save the Art World?* Beirut: Human & Urban and Alarm Editions.

Rabitti, P. (2008) *Ecoballe: Tutte le Verità su Discariche, Inceneritori, Smaltimento Abusive dei Rifiuti.* Rome: Aliberti.

Rancière, J. (2004) *The Politics of Aesthetics*, trans. G. Rockhill. New York, NY and London: Bloomsbury.

Rancière, J. (2010a) *Dissensus: On Politics and Aesthetics.* London and New York, NY: Continuum.

Rancière, J. (2010b) 'Communism: From Actuality to Inactuality', in *Dissensus: On Politics and Aesthetics.* London and New York, NY: Continuum, 76–83.

Rathje, W. and C. Murphy (2001) *Rubbish! The Archeology of Garbage.* Tucson, AZ: University of Arizona Press.

Reed, A., Jr (1999) 'The Black Urban Regime: Structural Origins and Constraints', in A. Reed, Jr, *Stirrings in the Jug: Black Politics in the Post-Segregation Era.* Minneapolis, MN: University of Minnesota Press.

Reno, J. (2014) 'Toward a New Theory of Waste: From "Matter Out of Place" to Signs of Life'. *Theory, Culture & Society* 31(6): 3–27.

Rice-Oxley, M. (2008) 'What's the Point of Recycling if It Just Goes to Waste?' *Guardian*, 15 January, available at: www.theguardian.com/environment/ethicalliving-blog/2008/jan/15/whatsthepointofrecyclingi (accessed 4 September 2015).

Rink, D. (2009) 'Wilderness: The Nature of Urban Shrinkage? The Debate on Urban Restructuring and Restoration in Eastern Germany'. *Nature and Culture* 4(3): 275–92.

Rittel, H.W.J. and M.M. Webber (1973) 'Dilemmas in a General Theory of Planning'. *Policy Sciences* 4(2): 155–69.

Robinson, J. (2006) *Ordinary Cities: Between Modernity and Development*. London: Routledge.

Robinson, J. (2014) 'Ebola Epidemic Spreads to Fifth West African Country as Case of Deadly Virus is Reported in Senegal – But Quarantine is Lifted in Slum Area of Liberian Capital'. *Daily Mail*, 30 August, available at: www.dailymail.co.uk/news/article-2738473/Ebola-epidemic-spreads-FIFTH-West-African-country-case-deadly-virus-reported-Senegal-quarantine-lifted-slum-area-Liberian-capital.html (accessed 8 October 2014).

Robinson, S.L. and J.T. Lundholm (2012) 'Ecosystem Services Provided by Urban Spontaneous Vegetation'. *Urban Ecosystems* 15(3): 545–57.

Rogers, H. (2005) *Gone Tomorrow. The Hidden Life of Garbage*. New York, NY: The New Press.

Rosa, B. (2014) 'Beneath the Arches: Re-appropriating the Spaces of Infrastructure in Manchester'. PhD dissertation. Manchester: University of Manchester.

Rose, G. (2012) 'The Question of Method: Practice, Reflexivity and Critique in Visual Culture Studies', in I. Heywood and B. Sandywell (eds), *The Handbook of Visual Culture*. London: Bloomsbury, 542–58.

Royte, E. (2005) *Garbage Land: On the Secret Trail of Trash*. New York, NY: Back Bay Books.

RSA (Royal Society of Arts) Nature Conservancy (1965) 'Reclamation and Clearance of Derelict Land'. London and Arlington, VA: RSA and Nature Conservancy.

Sabounchi, A. (2010) '"Ceci n'est Pas la Suisse": A Review Of Ayman Baalbaki's Solo Show at the Rose Issa Projects Gallery'. *Muraqqa*, March 2010, available at: www.muraqqa.com/ayman-baalbaki.html (accessed 13 November 2014).

Sader, H. (1998) 'Ancient Beirut: Urban Growth in the Light of Recent Excavations', in P. Rowe and H. Sarkis (eds), *Projecting Beirut: Episodes in the Construction and Reconstruction of a Modern City*. Munich: Prestel Verlag, 23–40.

Said, E. (1978) *Orientalism*. London: Penguin.

Sales, I. (1993) *La Camorra le Camorra*. Rome: Riuniti.

Salford City Council (2008) 'Salford Quays Milestones: The Story of Salford Quays'. Salford: Salford City Council.

Salhab, G. (dir.) (2007) *(posthume)*, video 28′.

San Francisco Environment (n.d.) 'Zero Waste: Nothing to Landfill in a Foreseeable Future', available at: www.sfenvironment.org/zero-waste (accessed 14 April 2015).

Santos, D. and J. Fux (2011) '*Estamira* e *Lixo Extraordinário*: a Arte na Terra Desolada'. *Ipotesi: Revista de Estudios Literários*, 15(2): 125–37.

Sassen, S. (1991) *The Global City: London, New York, Tokyo*. Princeton, NJ: Princeton University Press.

Saviano, R. (2008) *Gomorrah: Italy's Other Mafia*, London: Pan Books.

Scanlan, J. (2005) *On Garbage*. London: Reaktion Books.

Scanlan, J. (2013) 'Introduction: Aesthetic Fatigue, Modernity and Waste', in J. Scanlan and J.F.M. Clark (eds), *Aesthetic Fatigue: Modernity and the Language of Waste*. Newcastle upon Tyne: Cambridge Scholars Publishing.

Scharl, A. and K. Tochtermann (2007) *The Geospatial Web: How Geobrowsers, Social Software and the Web 2.0 are Shaping the Network Society*. London: Springer.

Schult, HA (1999) 'Trash People'. *HA Schult*, available at: www.haschult.de/action/trashpeople#content (accessed 4 September 2015).

Scott, J.C. (1990) *Domination and the Arts of Resistance: Hidden Transcripts*. New Haven, CT: Yale University Press.

Seale, K. (2005) 'Iain Sinclair's Excremental Narratives'. *M/C Journal*, 8(1), available at: http://journal.media-culture.org.au/0502/03-seale.php (accessed 11 June 2014).

Seale, K. (2006) 'Location, Location: Situating Bondi's "Rubbish House"'. *M/C Journal* 9(5), available at: http://journal.media-culture.org.au/0610/07-seale.php (accessed 7 September 2015).

Seale, K. (2014) 'On the Beach: Informal Street Vendors and Place in Copacabana and Ipanema, Rio de Janeiro', in C. Evers and K. Seale (eds), *Informal Urban Street Markets: International Perspectives*. New York, NY: Routledge.

Seigneurie, K. (2008) 'Anointing with Rubble: Ruins in the Lebanese War Novel'. *Comparative Studies of South Asia, Africa and the Middle East* 28(1), 50–60.

Senior, K. and A. Mazza (2004) 'Italian "Triangle of death" Linked to Waste Crisis'. *The Lancet Oncology* 5(9): 525–7.

Sennett, R. (2008) *The Craftsman*. New Haven, CT and London: University of Yale Press.

Serres, M. (2011) *Malfeasance*. Stanford, CA: Stanford University Press.

Sganzerla, R. (dir.) (1968) *O Bandido da Luz Vermelha*.

Shanks, M., D. Platt, and W. Rathje (2004) 'The Perfume of Garbage: Modernity and the Archaeological'. *Modernism/Modernity* 11(1): 61–87.

Sharpley, R. and P.R. Stone (eds) (2009) *The Darker Side of Travel: The Theory and Practice of Dark Tourism*. Bristol: Channel View Publications.

Sharpley, R. and P.R. Stone (eds) (2012) *Contemporary Tourist Experiences: Concepts and Consequences*. New York, NY: Routledge.

Shaviro, S. (2010) *Post-Cinematic Affect*. Ropley: Zero Books.

Sheridan, D. (2012) 'Disordering Public Space: Urban Wildscape Processes in Practice', in A. Jorgensen and R. Keenan (eds), *Urban Wildscapes*, Abingdon and New York, NY: Routledge, 201–20.

Sheringham, M. (2006) *Everyday Life: Theories and Practices from Surrealism to the Present*. Oxford: Oxford University Press.

Shields, R. (1991) *Places on the Margin: Alternative Geographies of Modernity*. London and New York, NY: Routledge.

Simone, A. (2004) *For the City Yet to Come: Changing African Life in Four Cities*. Durham, NC: Duke University Press.

Simone, A. (2005) 'Introduction', in A. Simone and A. Abouhani (eds), *Urban Africa: Changing Contours of Survival in the City*. Dakar: CODESRIA, 1–26.

Sinclair, I. (2003) *London Orbital*. Harmondsworth: Penguin.

Sinclair, I. (2004) *Dining on Stones*. Harmondsworth: Penguin.

Sinclair, I. (2005) *Edge of the Orison: In the Traces of John Clare's 'Journey Out of Essex'*. London: Penguin.

Sinclair, I. (2011) *Ghost Milk*. London: Hamish Hamilton.

Sinclair, I. and E. Matthews (2002) *White Goods*. Uppingham: Goldmark.

Sinn, E. (2008) 'Lesson in Openness: Creating a Space of Flow in Hong Kong', in H.F. Siu and A.S. Ku (eds), *Hong Kong Mobile: Making a Global Population*. Hong Kong: Hong Kong University Press, 13–44.

Siu, H.F. and A.S. Ku (eds) (2008) 'Introduction', in H.F. Sui and A.S. Ku (eds), *Hong Kong Mobile: Making a Global Population*. Hong Kong: Hong Kong University Press, 1–8.

Siu, H.F. and A.S. Ku (eds) (2008) 'Lessons in Openness: Hong Kong as a Space of Flow', in H.F. Sui and A.S. Ku (eds), *Hong Kong Mobile: Making a Global Population*. Hong Kong: Hong Kong University Press, 9–12.

Smart, A. and J. Lee (2009) 'Financialization and the Role of Real Estate in Hong Kong's Regime of Accumulation'. *Economic Geography* 79(2): 153–71.

Smith, N. (1984) *Uneven Development: Nature, Capital and the Production of Space.* Oxford: Blackwell.

Smith, V. (2007) *Clean: A History of Personal Hygiene and Purity.* Oxford: Oxford University Press.

Smoltczyk, A. (2008) 'Unsere Geschwüre seid ihr'. *Der Spiegel*, 14 January, available at: www.spiegel.de/spiegel/print/d-55410999.html (accessed 4 September 2015).

Sohn-Rethel, A. (1977) *Intellectual and Manual Labour: A Critique of Epistemology.* Atlantic Highlands, NJ: Humanities Press.

Soja, E. (2000) *Postmetropolis: Critical Studies of Cities and Regions.* Oxford: Blackwell.

Soja, E. (2011) *Postmodern Geographies: The Reassertion of Space in Critical Social Theory.* London: Verso.

Solà-Morales Rubió, I. de (1995) 'Terrain Vague', in C. Davidson (ed.), *Anyplace.* Cambridge, MA: MIT Press, 118–23.

Southworth, M. (2001) *Wastelands in the Evolving Metropolis.* Institute of Urban and Regional Development Working Paper Series, 2000-01. Berkeley, CA: University of California.

Spelman, E.V. (2011) 'Combing Through Trash: Philosophy Goes Rummaging'. *The Massachusetts Review* 52(2): 313–25.

Stallybrass, P. and A. White (1986) *The Politics and Poetics of Transgression.* Ithaca, NY: Cornell University Press.

Stallybrass, P. and A. White (2007) 'The City: The Sewer, the Gaze and the Contaminating Touch', in M. Lock and J. Farquhar (eds), *Beyond the Body Proper: Reading the Anthropology of Material Life.* Durham, NC: Duke University Press.

Stam, R. (1999) 'Palimpsestic Aesthetics: A Meditation on Hybridity and Garbage', in M. Joseph and J. Fink (eds), *Performing Hybridity.* Minneapolis, MN: University of Minnesota Press.

Stanley, H.M. (1890, 2 vols) *In Darkest Africa Or the Quest, Rescue, and Retreat of Emin Pasha, Governor of Equatoria.* New York, NY: Scribner.

Stavrides, S. (2014) 'Open Space Appropriations and the Potentialities of a "City of Thresholds"', in M. Mariani and P. Barron (eds), *Terrain Vague: Interstices at the Edge of the Pale.* Abingdon and New York, NY: Routledge, 48–61.

Stewart, S. (1994) *On Longing: Narratives of the Miniature, the Gigantic, the Souvenir, the Collection.* Durham, NC: Duke University Press.

Stoekl, A. (2007) *Bataille's Peak: Energy, Religion, and Postsustainability.* Twin Cities, MN: University of Minnesota Press.

Stoler, A.L. (2009) *Along the Archival Grain: Epistemic Anxieties and Colonial Common Sense.* Princeton, NJ: Princeton University Press.

Story, T.J. (2011) 'The Zero-Waste Home'. *Sunset*, available at: www.sunset.com/home/natural-home/zero-waste-home-0111/zero-waste-dining-room-0111 (accessed 30 September 2014).

Strangleman, T. (2013) '"Smokestack Nostalgia," "Ruin Porn" or Working-Class Obituary: The Role and Meaning of Deindustrial Representation'. *International Labor and Working-Class History*, 84: 23–37.

Strasser, S. (1999) *Waste and Want: A Social History of Trash*, New York, NY: Holt.

Stuart, A. (2013) 'As it Happened: Metrolink Director Answers Your Questions'. *Manchester Evening News*, 7 August, available at: www.manchestereveningnews.co.uk/

news/greater-manchester-news/live-metrolink-director-webchat-11am-5673628 (accessed 7 September 2015).

Sugrue, T. (2005) *The Origins of the Urban Crisis: Race and Inequality in Postwar Detroit.* Princeton, NJ: Princeton University Press.

Swyngedouw, E. (2004) 'Globalisation or Glocalisation? Networks, Territories and Rescaling'. *Cambridge Review of International Affairs* 17(1): 25–48.

Swyngedouw, E. (2009) 'The Antinomies of the Post-political City: In Search of a Democratic Politics of Environmental Production'. *International Journal of Urban and Regional Research* 33(3): 601–20.

Swyngedouw, E. (2010) 'Apocalypse Forever? Post-Political Populism and the Spectre of Climate Change'. *Theory, Culture & Society* 27(2/3): 213–32.

Symonds Travers Morgan (1996) 'Pomona Triangle, Cornbrook: Position Statement', Prepared for Central Manchester Development Corporation, Altrincham, Cheshire.

Synenko, J. (2014) 'The Remainders of Memory: Berlin's Postnational Aesthetic'. *Drain: A Journal of Contemporary Art and Culture*, 11(2), available at: www.academia. edu/9660183/The_Remainders_of_Memory_Berlins_Postnational_Aesthetic_In_ Drain_A_Journal_of_Contemporary_Art_and_Culture_Special_Issue_Ruin_Ed._ Ricky_Varghese_Vol._11_2_2014 (accessed 7 September 2015).

Tabet, R. (2012a) *1989.* Canvas and wooden frames.

Tabet, R. (2012b) *Architecture Lessons.* Concrete sculpture and video.

Tabet, R. (2013) 'If I Were to Write a Text on Sculpture'. *Peeping Tom Digest* 3 (Beirut), 76–7.

Tabet, R. and Greenberg, J. (2012) *The Ungovernables.* Short story with the installation *1989.*

Tamraz, N. (2009) 'Ayman Baalbaki's Mythological City', in G.H. Rabbath, *Can One Man Save the Art World?* Beirut: Human & Urban and Alarm Editions.

Tarasen, N. (2014) 'Double Negative: A Website about Michael Heizer', available at: http://doublenegative.tarasen.net/ (accessed 4 September 2015).

Taylor, A. (dir.) (2008) *Examined Life.*

Taylor, D. (2003) *The Archive and the Repertoire: Performing Cultural Memory in the Americas.* Durham, NC: Duke University Press.

Taylor, I., K. Evans, and P. Fraser (1996) *A Tale of Two Cities: Global Change, Local Feeling and Everyday Life in the North of England. A Study in Manchester and Sheffield.* London and New York, NY: Routledge.

Temple, J. (dir.) (2000) *The Filth and the Fury.*

TEP (2008) 'Toward a Green Infrastructure Framework for Greater Manchester: Full Report'. Warrington: TEP.

The Peel Group (2009) 'Ocean Gateway Prospectus, October 2009, Version 3'. Manchester: The Peel Group.

The Peel Group (2015) *Our Book.* Manchester: The Peel Group.

Thomas, J. (2013) *Redevelopment and Race: Planning A Finer City in Postwar Detroit.* Detroit, MI: Wayne State University Press.

Thompson, E. (2002) *The Soundscape of Modernity: Architectural Acoustics and the Culture of Listening in America, 1900–1933.* Cambridge, MA: MIT Press.

Thompson, H. (2004) *Whose Detroit? Politics, Labor, and Race in A Modern American City.* Ithaca, NY: Cornell.

Thomson, V.E. (2009) *Garbage In, Garbage Out. Solving the Problems with Long-Distance Trash Transport.* Charlottesville, VA: University of Virginia Press.

Tocqueville, A. de (1956) *Journeys to England and Ireland*. New Haven, CT and London: Yale University Press.

Tonkiss, F. (2013) 'Austerity Urbanism and the Makeshift City'. *City* 17(3): 312–24.

Tonnelat, S. (2008) ' "Out of frame": The (In)visible Life of Urban Interstices – A Case Study in Charenton-le-Pont, Paris, France'. *Ethnography* 9(3): 291–324.

Townsend, C.A. (2003) 'Chemicals From Coal: A History of Beckton Products Works', available at: www.glias.org.uk/Chemicals_from_Coal/INDEX.HTM (accessed 5 June 2014).

Trancik, R. (1986) *Finding Lost Space: Theories of Urban Design.* New York, NY: Van Nostrand Reinhold Company.

Trotter, D. (1988) *Circulation: Defoe, Dickens and the Economies of the Novel.* Basingstoke: Macmillan.

Trotter, D. (2000) *Cooking with Mud: The Idea of Mess in Nineteenth-Century Art and Fiction.* Oxford: Oxford University Press.

Turner, E. (1989) 'Report for Central Manchester Development Corporation: Land Use in 1986 in C.M.D.C. Area in the Context of City Centre Land Use as a Whole'. Manchester: City of Manchester Economic Development Department.

UNHabitat (United Nations Human Settlements Programme) (2010) *Solid Waste Management in the World's Cities: Water and Sanitation in the World's Cities 2010.* London and Washington, DC: UN-HABITAT/Earthscan.

United Nations (2014) 'World's Population Increasingly Urban With More Than Half Living in Urban Areas', available at: www.un.org/en/development/desa/news/population/world-urbanization-prospects-2014.html (accessed 14 April 2015).

Urrea, A. (1996) *By the Lake of Sleeping Children*. Harpswell: Anchor.

USHMM (United States Holocaust Memorial Museum) (2007) 'United States Holocaust Memorial Museum and Google Join in Online Darfur Mapping Initiative', available at: www.ushmm.org/information/press/press-releases/united-states-holocaust-memorial-museum-and-google-join-in-online-darfur-ma (accessed 4 September 2015).

Van Dijk, H. (1996) 'Colonizing the Void'. *A + U*. Tokio: Shinkenchiku-sha, 76–7.

Van Reybrouck, D. (2014) *Congo: The Epic History of a People*. London: Fourth Estate.

Varda, A. (dir). (2002) *The Gleaners and I*.

Vaz, C. (2009) 'My Dream of a "Zero Waste" Goa'. *Guardian Weekly*, 30 November, available at: www.theguardian.com/world/2009/nov/30/india-waste (accessed 4 September 2015).

Wacquant, L. (2008) *Urban Outcasts: A Comparative Sociology of Advanced Marginality*. Cambridge: Polity.

Walker, L. (dir.) (2010) *Waste Land*.

Walsh, E. (1975) *Dirty Work, Race, and Self-Esteem*. Ann Arbor, MI: The Institute of Labor and Industrial Relations; University of Michigan, Wayne State University.

Walsh, J. (2011) 'Pomona Docks: The Case for SBI Status'. Salford: Greater Manchester Ecology Unit.

Ward, K. (2003) 'Entrepreneurial Urbanism, State Restructuring and Civilizing "New" East Manchester'. *Area* 35(2): 116–27.

Watts, J. (2009) 'Chinese Protesters Confront Police Over Incinerator Plans in Guangzhou'. *Guardian*, 23 November, available at: www.theguardian.com/environment/2009/nov/23/china-protest-incinerator-guangzhou (accessed 4 September 2015).

White, L. and E. Frew (eds) (2013) *Dark Tourism and Place Identity: Managing and Interpreting Dark Places.* New York, NY: Routledge.

Whiteley, G. (2011) *Junk: Art and the Politics of Trash*. London and New York, NY: IB Tauris.

Whitford, J. (1967 [1877]) *Trading Life in Western and Central Africa*. London: Frank Cass.

Williams, J. (2014) '"Intimidating" Wasteland Around Cornbrook Metrolink Tram Stop to be Transformed into a Hotel, Office and Residential Hub'. *Manchester Evening News*, 2 September, available at: www.manchestereveningnews.co.uk/news/greater-manchester-news/intimidating-wasteland-around-cornbrook-metrolink-7700835 (accessed 7 September 2015).

Wilson-Goldie, K. (2006) 'Taking a Sense of Place – and Moving It'. *Daily Star*, 14 June, available at: www.dailystar.com.lb/Culture/Art/2006/Jun-14/112867-taking-a-sense-of-place-and-moving-it.ashx#ixzz3IwRLBKkz (accessed 13 November 2014).

Wilson-Goldie, K. (2007) 'Lebanon Bans Tale of Fighters in Militias'. *New York Times*, 18 August, available at: www.nytimes.com/2007/08/18/theater/18perf.html (accessed 20 July 2012).

Yassin, R. (2012a) *China*. Seven Porcelain Vases. Variable Dimensions.

Yassin, R. (2012b) '"China" Interview', in N. Muller (ed.), *Spectral Imprints*. Dubai: Abraaj Capital Art Prize, 65–92.

Yates, M. (2011) 'The Human-as-Waste'. *Antipode* 43(5): 1679–95.

Yoshida, R. (2009) '"Warship Island" Open to Tourists Once More'. *Japan Times*, 23 April, available at: www.japantimes.co.jp/news/2009/04/23/national/warship-island-open-to-tourists-once-more/#.Ve2ZE_STXgo (accessed 7 September 2015).

Young, J.E. (1994) *The Texture of Memory: Holocaust Memorials and Meaning*. New Haven, CT: Yale University Press.

Young, J.E. (2002) *At Memory's Edge: After-Images of the Holocaust in Contemporary Art and Architecture*. New Haven, CT: Yale University Press.

ZA Architects (2011) 'Revitalization of the Chernobyl Zone', available at: www.zaarchitects.com/en/projects/1/78-chernobyl (accessed 4 September 2015).

Zero Waste Detroit (2014) 'A Coalition for Recycling and an End to Waste Incineration', available at: http://zerowastedetroit.org/ (accessed 4 September 2015).

Žižek, S. (2006) *The Parallax View*. Cambridge, MA: The MIT Press.

Žižek, S. (2008) *The Sublime Object of Ideology*. London and New York, NY: Verso.

Zorach, R. (2005) *Blood, Milk, Ink, Gold*. Chicago, IL: University of Chicago Press.

Zukin, S. (1995) *The Cultures of Cities*. Cambridge, MA and Oxford: Blackwell.

Zurier, R. (2006) *Picturing the City: Urban Vision and the Ashcan School*. Berkeley, CA: University of California Press.

Index

Page numbers in *italics* denote tables, those in **bold** denote figures.

Milton Keynes UK
Ingram Content Group UK Ltd.
UKHW031145141024
449569UK00024B/1047